ATTRACTORS FOR SEMIGROUPS AND EVOLUTION EQUATIONS

In this volume, Olga A. Ladyzhenskaya expands on her highly successful 1991 Accademia Nazionale dei Lincei lectures. The lectures were devoted to questions of the behaviour of trajectories for semigroups of nonlinear bounded continuous operators in a locally non-compact metric space and for solutions of abstract evolution equations. The latter contain many initial boundary value problems for dissipative partial differential equations. This work, for which Ladyzhenskaya was awarded the Russian Academy of Sciences' Kovalevskaya Prize, reflects the high calibre of her lectures; it is essential reading for anyone interested in her approach to partial differential equations and dynamical systems.

This edition, reissued for her centenary, includes a new technical introduction, written by Gregory A. Seregin, Varga K. Kalantarov and Sergey V. Zelik, surveying Ladyzhenskaya's works in the field and subsequent developments influenced by her results.

Olga A. Ladyzhenskaya was a prolific Russian mathematician most well known for her work on partial differential equations and fluid mechanics. She authored over 200 hundred research works and became Head of the Mathematical Physics Laboratory of the Steklov Institute in 1961. Her many accolades include giving the Emmy Noether lecture at the International Congress of Mathematicians in 1994; giving the von Neumann lecture, the highest distinction of the Society for Industrial and Applied Mathematics, in 1998; and the Lomonosov Gold Medal, the highest award of the Russian Academy of Sciences, in 2002.

CAMBRIDGE
MATHEMATICAL LIBRARY

Cambridge University Press has a long and honourable history of publishing in mathematics and counts many classics of the mathematical literature within its list. Some of these titles have been out of print for many years now and yet the methods which they espouse are still of considerable relevance today.

The *Cambridge Mathematical Library* provides an inexpensive edition of these titles in a durable paperback format and at a price that will make the books attractive to individuals wishing to add them to their own personal libraries. Certain volumes in the series have a foreword, written by a leading expert in the subject, which places the title in its historical and mathematical context.

A complete list of books in the series can be found at www.cambridge.org/mathematics. Recent titles include the following:

Attractors for Semigroups and Evolution Equations
OLGA A. LADYZHENSKAYA

Fourier Analysis
T. W. KÖRNER

Transcendental Number Theory
ALAN BAKER

An Introduction to Symbolic Dynamics and Coding (Second Edition)
DOUGLAS LIND & BRIAN MARCUS

Reversibility and Stochastic Networks
F. P. KELLY

The Geometry of Moduli Spaces of Sheaves (Second Edition)
DANIEL HUYBRECHTS & MANFRED LEHN

Smooth Compactifications of Locally Symmetric Varieties (Second Edition)
AVNER ASH, DAVID MUMFORD, MICHAEL RAPOPORT & YUNG-SHENG TAI

Markov Chains and Stochastic Stability (Second Edition)
SEAN MEYN & RICHARD L. TWEEDIE

ATTRACTORS FOR SEMIGROUPS AND EVOLUTION EQUATIONS

OLGA A. LADYZHENSKAYA

With an Introduction by

GREGORY A. SEREGIN
Steklov Institute of Mathematics, St Petersburg and University of Oxford

VARGA K. KALANTAROV
Koç University, Istanbul

SERGEY V. ZELIK
University of Surrey

CAMBRIDGE
UNIVERSITY PRESS

CAMBRIDGE
UNIVERSITY PRESS

University Printing House, Cambridge CB2 8BS, United Kingdom

One Liberty Plaza, 20th Floor, New York, NY 10006, USA

477 Williamstown Road, Port Melbourne, VIC 3207, Australia

314–321, 3rd Floor, Plot 3, Splendor Forum, Jasola District Centre, New Delhi – 110025, India

103 Penang Road, #05–06/07, Visioncrest Commercial, Singapore 238467

Cambridge University Press is part of the University of Cambridge.

It furthers the University's mission by disseminating knowledge in the pursuit of education, learning, and research at the highest international levels of excellence.

www.cambridge.org
Information on this title: www.cambridge.org/9781009229821
DOI: 10.1017/9781009229814
Originally published in the Lezioni Lincee series

First published 1991
Re-issued 2008
Reprinted with Introduction 2022

A catalogue record for this publication is available from the British Library.

ISBN 978-1-009-22982-1 Paperback

Contents

Contents

Introduction

Gregory A. Seregin, Varga K. Kalantarov and Sergey V. Zelik

The year 2022 is the 100th anniversary of Olga A. Ladyzhenskaya, a famous Russian mathematician who played an outstanding role in developing the modern theory of partial differential equations (PDEs) and their applications including the qualitative theory of PDEs, infinite-dimensional dynamical systems, mathematical problems of hydrodynamics and nonlinear problems of mathematical physics.

The book we are presenting is the expanded version of Ladyzhenskaya's lecture notes for the course of lectures given by her at the Accademia Nazionale dei Lincei in 1991 and the aim of this book is to give a brief introduction to the mathematical foundations of the theory of infinite-dimensional dynamical systems and their attractors, with applications to several classes of dissipative nonlinear PDEs.

Ladyzhenskaya came to attractors from her favourite topic, the mathematical theory of viscous incompressible fluids, by trying to understand the nature of turbulence by interpreting the Navier–Stokes system as an infinite-dimensional dynamical system and using/extending the ideas and methods of classical dynamics. Such an interpretation became possible due to her fundamental result on the uniqueness of solutions for the 2D Navier–Stokes problem, proven in 1958, see [62] (see also [63] for a more detailed exposition). Precisely, it has been proved there that the initial boundary value problem

$$\begin{cases} \partial_t v + \sum_{k=1}^{2} v_k v_{x_k} + \nabla p = \nu \Delta v + f, \quad \operatorname{div} v = 0, \quad t > 0,\ x \in \Omega, \\ v = 0, \quad x \in \partial\Omega, \quad v(x,0) = v_0(x),\ x \in \Omega, \end{cases} \tag{1}$$

in a bounded domain $\Omega \subset \mathbb{R}^2$ possesses a unique global solution $v(t)$ for all external forces f and initial conditions v_0 belonging to a properly chosen function space.

When this global well-posedness is established one can define a solution semigroup V_t of this problem in the proper phase space H and treat the Navier–Stokes equations as an infinite-dimensional *dynamical system* V_t acting on this phase space H (analogously with the classical qualitative theory of dynamical systems generated by ODEs). In her seminal paper [65], Ladyzhenskaya constructs a special set $\mathcal{M}$ in the phase space H (roughly speaking, the set of square integrable solenoidal vector fields) with the following properties:

- $\mathcal{M}$ is *invariant*: $V_t(\mathcal{M}) = \mathcal{M}$, i.e. $\mathcal{M}$ consists of trajectories of the dynamical system V_t;
- all trajectories are *attracted* by $\mathcal{M}$, i.e. any trajectory that started from a bounded set B comes to an ε-neighbourhood of $\mathcal{M}$ after a finite time $T(\varepsilon, B)$ and remains there;
- $\mathcal{M}$ is *compact* in H and is therefore in a sense finite-dimensional.

This set is constructed as an ω-limit set of the absorbing ball of the semigroup V_t and is *exactly* what is nowadays called *a global attractor*.

> Ladyzhenskaya wrote in [69] about the reasons that prompted her to study attractors: I tried to understand what the experimenter can observe after a very long (infinite) period of time. At the same time, I started from the statement widespread among physicists that the solutions of dissipative systems 'forget' their initial data and are 'formed' under the influence of constantly (stationary) acting factors. In the literal sense of the word, this, of course, is not true, because in a deterministic system (I had in mind only such systems, and first of all the two-dimensional Navier–Stokes equations, for which the global unique solvability of the initial-boundary value problems has been proved) solutions are determined by their initial data (as well as boundary conditions and external forces, which are considered fixed and independent of time). But in the course of time, the solution may move far away from them and, in this sense, forget them. And I asked myself the question: what is the part of the phase space to which solutions are attracted and what is the dynamics on this part?

These words are very similar to the modern description of so-called *deterministic chaos*, an extremely interesting and important, and in a sense still 'mysterious' phenomenon which allows a deterministic system to demonstrate random behaviour. Note that this phenomenon has been observed in hydrodynamics and weather prediction by Lorentz [78] and is nowadays considered to be one of the characteristic features of turbulence.

The prominent ideas of Ladyzhenskaya inspired many brilliant mathematicians to switch to this area and led to the development of a general theory of infinite-dimensional dynamical systems governed by dissipative PDEs (also known as *attractor theory*), which is the main subject of the book we are presenting.

The book consists of two parts. Part I, comprising Chapters 1–4, is focused on the theory of global attractors for semigroups defined on a complete metric space X. Part II containing Chapters 5–7 is about particular semigroups generated by initial boundary value problems for 2D Navier–Stokes equations, nonlinear parabolic equations, damped nonlinear wave equations, etc.

In Chapter 1, a number of basic notions are defined. Since some of them differ from the modern terminology, we reproduce them and indicate differences.

- A semigroup $V_t\colon X \to X$ is called a *bounded semigroup* if for each bounded set $B \subset X$, the orbit set $\gamma^+(B) := \cup_{x\in B}\{V_t x,\ t \in \mathbb{R}^+\}$ is bounded.
- A set $A \subset X$ *attracts* a set $M \in X$ if for each $\varepsilon > 0$ there exists $T = T(\varepsilon, M)$ such that $V_t(M) \subset O_\varepsilon(A)$ for all $t \geq T$, where $O_\varepsilon(A)$ is an ε-neighbourhood of the set A.
- A set A attracting each point of X is called a *global attractor* of the semigroup V_t and a set that attracts each bounded set $B \subset X$ is called the *global B-attractor* of the semigroup V_t.
- A semigroup is called *pointwise dissipative* if it has a bounded global attractor and it is called *B-dissipative* if it has a bounded global B-attractor.
- A ball B_R is called an *absorbing* set of the semigroup V_t if for each bounded set $B \subset X$ there exists $t_0(B)$ such that $V_t(B) \subset B_R$, $\forall t \geq t_0(B)$.

In modern terminology, a semigroup or dynamical system is called *dissipative* if it possesses a bounded absorbing ball. Moreover, a global B-attractor is nowadays simply known as a global attractor, and Ladyzhenskaya's 'global attractor' is known as a pointwise attractor. Thus, on the one hand, the attractor $\mathcal{M}$ captures all of the non-trivial limit dynamics of the system in question as time goes to infinity, while, on the other hand, it is essentially smaller than the initial phase space. In particular, in many cases, this attractor has a finite dimension, so there is a tremendous reduction of the effective 'degrees of freedom' (from infinite to finite), which in turn allows us to use the ideas and methods of classical dynamics to investigate the dynamics of PDEs.

The main objects of study in Chapter 2 are so-called *semigroups of class $\mathcal{K}$*, i.e. semigroups that for each $t > 0$ map any bounded set $B \subset X$ to a precompact set $V_t(B)$. Two important statements, Theorem 2.2 and Theorem 2.3, are proved in the chapter. Conceptually, they show the following:

- if a continuous semigroup $V_t : X \to X$ is of class $\mathcal{K}$ and B-dissipative or bounded and point-wise dissipative, then it possesses a minimal global B-attractor $\mathcal{M}$ which is compact, invariant and connected, provided X is connected;
- if $V_t : X \to X$ is a continuous semigroup of class $\mathcal{K}$, the orbit $\gamma^+(x)$ is bounded for each $x \in X$ and a 'good' Lyapunov function exists, then there is a minimal global attractor $\widehat{\mathcal{M}}$ that consists of the stationary points Z of the semigroup. In other words, as time tends to infinity any bounded trajectory of the semigroup converges to the set Z. Moreover, if Z is bounded, then the semigroup has a minimal global B-attractor $\mathcal{M}$. In addition, if Z is totally disconnected (e.g. if it consists of finitely many points), then the attractor $\mathcal{M}$ consists of equilibria Z and all heteroclinic orbits connecting different equilibria from Z.

Of course, there are many publications (see e.g. [6, 42, 44, 45, 85, 91] and references therein) where various theorems about attractors of semigroups of class $\mathcal{K}$ and/or their applications have been proven. To our knowledge, the first result on attractors of compact semigroups appeared in [11] and the results for the case where the Lyapunov function exists can be found in [42].

It is noteworthy that the class of semigroups possessing a global Lyapunov function is extremely important for the theory of attractors since it is the only known relatively large class for which we can say something reasonable about the structure of the attractor. The further development of this theory led to the concept of a *regular* attractor: a global attractor that consists of a finite union of finite-dimensional unstable manifolds of equilibria, see [8] and also [92] for the modern state of the art. Such attractors have a lot of good properties which usually fail for general attractors. For instance, they are robust with respect to perturbations and the rate of attraction to them of bounded sets is exponential. Moreover, under the 'generic' assumption that stable and unstable manifolds of equilibria intersect transversally (the Morse–Smale property), the dynamics on these attractors is also robust with respect to perturbations, see [17, 18, 33]. We also mention that the assumption that the equilibria set Z is totally disconnected can be partially removed using the Simon–Lojasiewicz technique, which gives the stabilization to a *single* equilibrium even in the case that Z is a continuous set, see e.g. [53].

The results of Chapter 2 are extended in the subsequent Chapter 3 to the class of *asymptotically compact* semigroups (semigroups of class $\mathcal{AK}$). These semigroups possess the property that each bounded sequence $V_{t_k}(x_k)$, with $\{x_k\} \subset X$ and $t_k \to \infty$, is a precompact set of X. As shown in Chapter 3, the main results proved for semigroups of class $\mathcal{K}$ remain valid for $\mathcal{AK}$-semigroups.

Asymptotically compact semigroups arise naturally in the study of non-parabolic equations (e.g. damped wave equations) that do not have the simultaneous smoothing property. Theorem 3.3 gives the main technical tool to study such equations. It claims that a semigroup V_t is asymptotically compact if it can be presented as a sum

$$V_t = U_t + W_t, \tag{2}$$

where the operators U_t are compact for every fixed t and W_t tend to zero uniformly with respect to bounded sets as time tends to infinity.

We mention that the theory of $\mathcal{AK}$ semigroups has undergone intense development during the last two decades and nowadays we have a number of effective methods for verifying their asymptotic compactness which do not require the splitting of the semigroup: for instance, the so-called *energy* method (see [10, 84]) or the methods based on compensated compactness or precompact semi-norms (see [24]).

Chapter 4 is devoted to upper bounds for the Hausdorff and fractal dimension of the attractors. The main result here is the proof of the classical volume contraction theorem, which states that a C^1-map V on a Hilbert space H contracts N-dimension volume in some neighbourhood of a compact invariant set $\mathcal{A}$, then the Hausdorff dimension of $\mathcal{A}$ does not exceed N, see Theorem 4.5. Combined with the Liouville formula for the evolution of k-dimensional volumes (see formula (4.28)) this yields one of the most popular modern methods for estimating the dimension of the attractors; it is especially effective for hydrodynamical problems. The analogous result for the fractal (box-counting) dimension is also given (see Theorem 4.6), but the estimate is essentially weaker and is not so elegant.

Note that while this key theorem for the Hausdorff dimension was obtained for the finite-dimensional case in [52] and for the infinite-dimensional case in [27], the same result for the fractal dimension was open for a long time and has only recently been established. A breakthrough on this problem came in the paper [49] where the result was obtained in the finite-dimensional case. Then it was extended to the infinite-dimensional diffeomorphisms in [13] and the final result in exactly the same formulation as for the Hausdorff dimension was obtained in [22]. Thus, nowadays there is no difference in estimating Hausdorff and fractal dimension via the volume contraction method.

The second part of the book is devoted to applications of the general theory developed in Part I to the classical equations of mathematical physics, namely, the 2D Navier–Stokes equations (Chapter 6) and damped wave equations (Chapter 7).

Chapter 6 contains two main results (Theorems 6.1 and 6.2) which give the upper bounds for the number of determining modes N and the fractal dimension $\dim_f(\mathcal{M})$ of the attractor of the 2D Navier–Stokes equation (1) in terms of the parameter ν. The estimates obtained for N are:

$$N \leq c_1\nu^{-4} + c_1' \quad \text{and} \quad N \leq c_2\nu^{-2}|\ln\frac{1}{\nu}| + c_2'$$

for the cases of no-slip (Dirichlet) and periodic boundary conditions, respectively. Upper bounds for the fractal dimension are obtained for the case of no-slip boundary conditions only and have the same form as for the corresponding determining modes:

$$\dim_H(\mathcal{M}) \leq c_3\nu^{-4} + c_3',$$

where c_i and c_i' above are some constants that are independent of ν.

Here we would like to make some important remarks. There is a heuristic conjecture (partially inspired by the conventional theory of turbulence of A.N. Kolmogorov from the one side and I. Prigogine's theory of dissipative structures from the other) that despite the infinite-dimensionality of the initial phase space, the limit dynamics of a dissipative system are finite-dimensional and can be effectively described by the evolution of finitely many parameters (the so-called order parameters in the terminology of I. Prigogine). One of the ultimate goals of the theory of attractors is to find a rigorous interpretation and justification of this conjecture. Historically, the first attempt at tackling this problem was made in the pioneering works of Foias and Prodi [34] and of Ladyzhenskaya [65] using precisely the method of determining modes.

To be more precise, it was proved that the limit dynamics of 2D Navier–Stokes equations are determined in a unique way if the evolution in time of the first N Fourier modes is known and if N is large enough. So in some sense these limit dynamics are determined by N parameters.

The further development of this theory went in the direction of generalizing the form of the determining modes and computing upper bounds that are as sharp as possible for the number N. In particular, analogous results have been obtained in [36] where Fourier modes are replaced by *nodes* (i.e. the values of the dependent variable at the nodes of some spatial grid). Later on, the notions of determining volume elements and so on were introduced and various upper bounds for the number N of determining elements in such systems were obtained for various dissipative PDEs (see e.g. [36, 38, 39] and references therein). The more general notion of determining functionals (or determining interpolant operators) as well as a unified approach for investigating parameters uniquely determining asymptotic behaviour of solutions to dissipative PDEs was introduced in [26] and developed in [23] (see e.g. [24] and references therein).

The key drawback of the described approach to the problem of finite-dimensionality is that the values of the 'slaved' higher modes at some fixed time t cannot be found in terms of the values of determining functionals at the same moment in time (one needs to know the values of determining functionals *for all times* in order to do this). In other words, the time evolution of the values of determining functionals are not governed by a system of ODEs and still can be infinite-dimensional. A closely related example here is a system of ODEs with delay where the 'number of parameters' is finite, but the dynamics can still be infinite-dimensional. This drawback inspired researchers to seek a stronger version of finite-dimensional reduction based on various dimensions of the attractor $\mathcal{M}$.

To the best of our knowledge, the first result on the finiteness of the Hausdorff dimension of a negatively invariant set in a Hilbert space was obtained by J. Mallet-Paret in [79] with applications to delay differential equations as well as to the 1D reaction diffusion equation. Based on this result, the first very rough estimate for the Hausdorff dimension of a global attractor of the 2D Navier–Stokes equations was obtained in [35].

Later on, in [67], Ladyzhenskaya proved the following result, which can also be treated as a generalization of the method of [79].

Theorem A1 *Let M be a bounded subset of a Hilbert space H, and let $V: M \to H$ be an operator such that $M \subset V(M)$ and satisfying the conditions*

$$\|V(v) - V(w)\| \le \ell\|v - w\|, \ \forall v, w \in H,$$

and

$$\|Q_N(V(v) - V(w))\| \le \delta\|v - w\|, \ \forall v, w \in H,$$

where $\ell > 0$, $\delta \in (0,1)$ are given numbers and Q_N is the orthogonal projection onto the subspace of co-dimension N. Then

$$\dim_H(M) \le N \ln\left(\frac{8\kappa^2\ell^2}{1-\delta^2}\right)\left[\ln\frac{2}{1+\delta^2}\right]^{-1},$$

where $\kappa > 0$ is an absolute constant.

Using this theorem, she got an estimate of the Hausdorff dimension of the global attractor for the 2D Navier–Stokes equations that grows exponentially in ν^{-1} in the case of no-slip boundary conditions (see also [51]).

The polynomial in ν^{-1} estimate (6.14) presented in this monograph is due to Babin and Vishik [7] (for the case of Hausdorff dimension) and Constantin and Foias [27] (for fractal dimension). However, this estimate is still far from being optimal and further progress in this direction is due to the use of so-called Lieb–Thirring inequalities, see [75]. Up to the moment, the best known upper bounds for the fractal dimension of the 2D Navier–Stokes equations can be found in [91]:

$$\dim_f(\mathcal{M}) \le CG, \quad G := \|f\|_{L^2}|\Omega|\nu^{-2},$$

where the modern value of the constant C is related to the constant C_{LT} in the corresponding Lieb–Thirring inequality via $C = \frac{C_{LT}^{1/2}}{2\sqrt{2\pi}}$. The explicit value of this constant is not available, but the best known analytic bound is $C_{LT} \le \frac{1}{2\sqrt{3}}$, see [22, 40].

For the case of periodic boundary conditions, the obtained upper bounds can be essentially improved to:

$$\dim_f(\mathcal{M}) \le cG^{2/3}(1 + \log G)^{1/3},$$

see [91]. Moreover, this estimate is in a sense sharp up to the logarithmic term: the lower bounds of the form

$$\dim_f(\mathcal{M}) \ge c'G^{2/3}$$

are attained on the properly constructed Kolmogorov flows, see [77]. Note also that no non-trivial lower bounds are known for the case of no-slip boundary conditions.

It is also noteworthy that Theorem A1 has an essential advantage in comparison with other methods, namely, that the differentiability of the corresponding semigroup is not required. For this reason, it can be applied to many classes of degenerate or singular problems as well as problems with supercritical nonlinearities where this differentiability is problematic or is difficult to prove, see [24, 25, 82] and references therein for further generalizations and applications. In particular, this theorem is very useful for estimating the dimensions of attractors for various problems related to non-Newtonian fluids in dimensions two or three, see e.g. [66, 71–73] and [80].

We also mention that the proper generalizations of the Ladyzhenskaya squeezing property used in Theorem A1 have been exploited later in order to demonstrate the existence of the so-called exponential attractor for various dissipative nonlinear PDEs, see [82] and references therein. The notion of an exponential attractor, which is somehow an intermediate object between the global attractor and an inertial manifold, was introduced in [30] in order to overcome major drawbacks of global attractors (sensitivity to perturbations and slow rate of attraction). Remarkably, the initial assumptions of [30] for the existence of such an object are very close to the assumptions of Theorem A1.

As we have already mentioned, Chapter 7 of the book is devoted to attractors for abstract semilinear damped wave equations. After the short introduction, the exposition begins (in Section 7.2) with the detailed analysis of the linear problem

$$\partial_t^2 v + \nu \partial_t v + Av = h, \quad v(0) = v_0, \quad v_t(0) = v_1, \tag{3}$$

in an abstract Hilbert space H. Here $A\colon D(A) \to H$ is a given positive self-adjoint linear operator with compact inverse and $h = h(t)$ is a given external force which may depend explicitly on time. The main result of the section is the existence and uniqueness theorem for the solutions of problem (3) in the appropriate energy spaces, which nowadays has become a standard technical tool for the study of more general *nonlinear* wave equations.

We would like to emphasize here that Ladyzhenskaya was one of the first mathematicians who applied functional analytic methods to study the solvability of initial boundary value problems for hyperbolic equations, which nowadays is classical (including the results presented in Section 7.2). Her first book [61] deals exactly with equation (3) in the particular case that A is a second order symmetric uniformly elliptic differential operator in a bounded domain $\Omega \subset \mathbb{R}^n$.

Section 7.3 is devoted to the study of the analogous problems for the nonlinear wave equation of the form

$$\partial_t^2 v + \nu \partial_t v + Av + f(v) = 0, \quad v(0) = v_0, \quad v_t(0) = v_1, \tag{4}$$

where the nonlinear function f is a smooth enough gradient (i.e. $f(v) = \mathcal{F}'(v)$ for some given non-negative potential $\mathcal{F}$) and is in a sense subordinate to the leading linear part Av.

In the subsequent Section 7.4 Ladyzhenskaya studied the differentiability of solutions of problem (4) with respect to initial data for which the application of the volume contraction method and estimation of the fractal dimension of the corresponding attractor is necessary.

This attractor (a global B-attractor in the terminology of the book) is constructed in Section 7.5 by verifying the associated solution semigroup V_t in the standard energy space belongs to the class $\mathcal{AK}$. This fact, in turn, is obtained with the help of a decomposition as in (2) where U_t is the solution operator that corresponds to the *linear* problem (4) with $f \equiv 0$. Finally, the result about the finite-dimensionality of the corresponding attractor is proved in the concluding Section 7.6.

We recall that the key model for the abstract damped wave equation (4) is the following dissipative PDE:

$$\begin{cases} \partial_t^2 v + \nu \partial_t v - \Delta v + f(v) = h, \ v\big|_{\partial\Omega} = 0, \\ v(x,0) = v_0(x), \quad \partial_t v(x,0) = v_1(x), \end{cases} \tag{5}$$

where $\Omega \subset \mathbb{R}^n$ is a bounded domain with smooth boundary $\partial\Omega$, and $f \in C^1(\mathbb{R})$ satisfies the dissipativity conditions

$$f(s)s - \mathcal{F}(s) \geq -C, \ \mathcal{F}(s) := \int_0^s f(\tau)d\tau \geq -C, \ C > 0,$$

as well as the growth restriction

$$|f'(s)| \leq M(1 + |s|^{p-1}),$$

where $p \geq 1$ is a given growth exponent and $M > 0$.

In particular, the results presented in the book hold for this equation when

$$p < p_{en-crit} := \frac{2}{n-2} \ \ (p \text{ may be arbitrarily large if } n = 1 \text{ or } n = 2).$$

This is the so-called sub-critical energy case where the nonlinearity f is strongly subordinated to the Laplacian in the standard energy space. Actually, the results of Chapter 7 concerning global well-posedness hold for the energy critical case $p = p_{en-crit}$, but the method of verifying the asymptotic compactness requires $p < p_{en-crit}$. We emphasize that this damped wave equation is not the only equation to which the results of Chapter 7 can be applied. Among other interesting examples are various versions of nonlinear plate equations, von Karman equations, etc., see [24, 25] for more details.

It is noteworthy that the class of wave equations of the form (5) is one of the most important classes of PDEs and has been studied intensively by many prominent mathematicians. The first results on global existence, uniqueness, and regularity of weak solutions to the Cauchy problem for some cases of equation (5) with $p \in [1,4)$ (in 3D case) are demonstrated in [55] and for the initial boundary value problem with $p \in [1,3]$ in [76] and [88]. Later on, the above results were proved for the Cauchy problem with quintic nonlinearity ([58], [50]) and for the initial boundary value problem with $p \in [3,5]$ in [15]. The supercritical case $p > 5$ is much more delicate and the uniqueness of energy solutions in this case is still an open problem. The recent result of Tao

[90] shows that finite time blow-up of smooth solutions may appear in *systems* of equations of type (5) in the supercritical case.

With regards to the attractors for equation (5), a number of papers appeared on this topic in the beginning of the 1980s, although some preliminary results in this direction had been obtained a little earlier by Ball [9] and Webb [93].

The existence of a global attractor for equation (5) in the standard energy phase space in the energy subcritical case $p < p_{en-crit}$ was obtained by Haraux [46] and Hale [43], see also [70], and estimates for the dimension of this attractor were obtained by Ghidaglia and Temam in [41] and Ladyzhenskaya [68].

The energy critical case $p = p_{en-crit}$ has been treated by Babin and Vishik in [8] (see also Arieta, Carvalho and Hale in [2]); the existence of a global attractor for the critical case was also obtained by Ladyzhenskaya [70], but in higher energy space only. The key idea of their method is to use the so-called dissipation integral together with a slightly delicate decomposition as in (2) in which both operators U_t and W_t are nonlinear. This decomposition also allowed them to establish the extra smoothness of the constructed global attractor. Note that the usage of the dissipation integral was actually a serious restriction which did not allow them to extend the method to non-autonomous external forces or unbounded domains. This restriction was later removed in [94].

The existence of a global attractor for problem (5) on a compact n-dimensional Riemann manifold without boundary (e.g. on a torus that corresponds to periodic boundary conditions) was established in [59] (see also [32] for the analogous result for $\Omega = \mathbb{R}^3$) under the assumption

$$p \in (1, p_{\text{crit}}), \quad p_{\text{crit}} := \frac{4}{n-2}.$$

The key new idea that allows them to shift the limit exponent is related to the so-called Strichartz estimates, which give control of the $L^4(0,T;L^{12}(\Omega))$-norm of the solution and this in turn leads to the uniqueness of the properly defined weak solution. Remarkably, as is pointed out in [59], the idea of using Strichartz estimates in the theory of attractors actually came from Ladyzhenskaya.

In contrast to this, the analogous result *in a bounded domain* remained open for a long time, because of the absence of Strichartz estimates for manifolds with non-empty boundary. The breakthrough in the theory of such estimates was made in the mid-2000s, see [12, 15] and references therein. By combining these results with the classical Pohozhaev–Morawetz identity, the global well-posedness of (5) with the critical growth exponent $p=5$

(in 3D) was established for [15] for the class of so-called Shatah–Struwe solutions (i.e. energy solutions that have an extra space-time regularity $u \in L^4(0,T;L^{12}(\Omega))$).

The final result on the existence of a global attractor for (5) in a bounded domain for all growth exponents $1 \le p \le 5$ and its smoothness was established in [56] by combining Strichartz estimates and the Pohozhaev–Morawetz identity from the one side with the so-called backward regularity of weak solutions established earlier in [95] on the other side.

To conclude this exposition, we briefly mention several important branches of the modern theory of infinite-dimensional dynamical system that are not formally reflected in the book. Although some of these branches are formally not so close to her area of scientific interests, many fundamental results in these branches are actually inspired by the pioneering works of Ladyzhenskaya.

1) *Inertial manifolds (IMs).* By definition, these are finite-dimensional invariant sub-manifolds of the phase space that *exponentially* attract all trajectories of the considered dynamical system. An IM is a 'dream object' in the theory of infinite-dimensional dynamical systems. Indeed, it allows us not only to embed a global attractor into a finite-dimensional smooth (at least, $C^{1+\varepsilon}$-smooth) manifold, but also gives a very natural reduction of the limit dynamics to a system of ODEs simply by restricting the initial dynamics to the IM (this is the so-called Inertial Form of the initial PDE). Moreover, if an IM exists it is usually normally-hyperbolic, so it is robust with respect to perturbations and possesses the exponential tracking property.

This object was introduced in [37], inspired by the hope of understanding the nature of turbulence by reducing it to the study of a system of finitely many ODEs describing the evolution of the Prigogine order parameters (coordinates on the base of the manifold, which are very often just Fourier modes). Even the name 'inertial' is related to the inertial scale in turbulence and somehow reflects this hope.

Unfortunately, the existence of such a manifold requires strong extra assumptions on the considered system (the so-called spectral gap or cone conditions) which are not satisfied, e.g. the 2D Navier–Stokes equations, so the question of the existence or non-existence of IMs in turbulence has remained completely open for more than 30 years; already we refer the reader to the survey [97] for a modern exposition of this theory. We also mention that many new and interesting results concerning IMs for equations that do not satisfy the spectral gap conditions have appeared in the last few years, see [57] and references therein.

2) *Unbounded domains.* The theory of the attractors for PDEs in unbounded domains was initiated in the pioneering papers [1] and [4] and it has undergone

intense development from this time, see [3, 31, 32, 83, 87, 96] and also the survey [82] and references therein. The principal difference with the case of bounded domains (which has been observed already in [4]) is that the corresponding attractor may have and (typically has) an infinite dimension, so the finite-dimensional reduction no longer works and we are faced with a new type of dynamics that cannot be reduced to a finite system of ODEs. Some features of these dynamics are already understood nowadays (e.g. there are universal asymptotical formulas for Kolmogorov's ε-entropy that depend only on the shape of the domain and are independent of a concrete choice of the equation and even the type of this equation), which replace the estimates of fractal dimension discussed above, but many other features remain a mystery, see [82] for more details.

Another important difference is that the natural phase space for such problems naturally consists of *infinite-energy* solutions and this produces a lot of extra difficulties, especially for hydrodynamical problems, since we no longer have the cancellation of the inertial form, see [96] where a theory of attractors is built for the Cauchy problem for 2D Navier–Stokes equations with Ekman damping.

3) *Non-autonomous and random attractors.* There are at least two principally different ways to extend the concept of a global attractor to a non-autonomous system. The first treats the attractor as a time-independent set despite the considered system's explicit dependence on time, while the second treats the attractor as a time-dependent set as well. The first concept is related to the reduction of a non-autonomous dynamical system to an autonomous one and naturally leads to so-called *uniform* attractors (roughly speaking such an attractor attracts images of all bounded sets uniformly with respect to the initial moment in time, see [21] for more details). The second one leads to a time-dependent set $\mathcal{M} = \mathcal{M}(t)$ which usually fails to attract the trajectories forward in time and possesses the so-called *pullback* attraction property (roughly speaking, if you start with a bounded set very far in the past, the corresponding trajectory at a fixed present time t will be close to $\mathcal{M}(t)$), so they are usually referred to as *pullback* attractors (or *kernel sections* in the terminology of Vishik and Chepyzhov), see [16, 21, 60] for more details.

To the best of our knowledge, the first attempt at a systematic study of attractors for non-autonomous dynamical systems is due to Haraux [47, 48] where he introduced various types of non-autonomous attractors, studied relations between them and gave sufficient conditions for their existence (see also [11] and [9] for the cases of periodic and asymptotically autonomous dynamical systems, respectively). Attractors for random dynamical systems were introduced by Crauel and Flandoli [28] based on the idea of a pullback attractor, see also [20] for the deterministic counterpart of the theory of pullback attractors. The theory of uniform attractors has been developed intensively by Vishik and Chepyzhov since the mid-1990s and is summarized in their monograph [21]. We mention here only one of the results developed in the theory of nonautonomous attractors, which concerns their fractal dimension. Random and pullback attractors usually have finite fractal dimension, which can be estimated by a similar scheme as for the autonomous case. By contrast, uniform attractors are typically infinite-dimensional and dimension estimates for them can be naturally replaced with estimates for their Kolomogorov ϵ-entropy (similarly to the case of unbounded domains). We refer the interested reader to [16, 21, 60] and the references therein for more details.

4) *Equations without uniqueness and trajectory attractors.* It is clear that the uniqueness of a solution for the considered PDE is *crucial* for the theory built up in this book since it is used at the very first step in constructing an associated semigroup V_t. But what can we do for equations for which this uniqueness is unknown (e.g. for the 3D Navier–Stokes equations, where the existence, due to Leray [74], has been known since 1934 whereas the uniqueness is the Clay Mathematics Institute's sixth Millennium Prize Problem; we refer the reader to the book [86] and the papers [64], [54] and [14] for more details) or for equations with known non-uniqueness (e.g. elliptic boundary problems in cylindrical domains, see [5])? Is it possible to develop some kind of attractor theory for such problems?

The straightforward way to do this is to introduce, following [9] and [7], semigroups of multi-valued maps and to extend the theory of attractors to these semigroups, see also [5], [81] and [10]. However, there is another much more elegant and likely more effective way to handle the problem. Namely, instead of constructing the semigroup V_t on a classical phase space X of our problem via $V_t v_0 := v(t)$, we may consider the so-called *trajectory* phase space $\mathcal{K}_+$, which consists of all semitrajectories $v(t)$, $t \geq 0$ of our dynamical system (or some properly chosen shift invariant set of semi-trajectories) endowed with the appropriate topology (usually some local in time topology), and consider a semigroup $T(h)$ of time translations which acts on $\mathcal{K}_+$. We then refer to the semigroup $T(h)$ acting on $\mathcal{K}_+$ as a *trajectory dynamical system* related to our equation and study a global attractor $\mathcal{M}_{tr}$ of this semigroup, which is called a *trajectory attractor* for our equation. It can be shown that in the case where the uniqueness theorem holds, this construction gives a semigroup that is topologically conjugate to the standard semigroup V_t, so their attractors exist or do not exist simultaneously and $\mathcal{M} = \mathcal{M}_{tr}\big|_{t=0}$. The advantage is that the trajectory dynamical system does not require the uniqueness of a solution (only the existence of a solution is necessary to ensure that $\mathcal{K}_+$ is not empty).

Thus, the described construction allows us to study the long-term behaviour of equations without uniqueness without inventing new types of attractors, see [21, 82] and references therein for more details.

The trajectory attractors were introduced by Chepyzhov and Vishik [19] and by Sell [89], motivated mainly by applications to the 3D Navier–Stokes equations, see [21, 82] and references therein for more details, and have nowadays become one of the most powerful objects for investigating the long-term behaviour of solutions of PDEs without uniqueness.

References

[1] F. Abergel (1990). Existence and finite dimensionality of the global attractor for evolution equations on unbounded domains. *J. Differ. Equ.* **83**:1, 85–108.

[2] J. Arrieta, A.N. Carvalho and J. K. Hale (1992). A damped hyperbolic equation with critical exponent. *Commun. Partial. Differ. Equ.* **17**:5-6, 841–866.

[3] A.V. Babin (1992). The attractor of a Navier-Stokes system in an unbounded channel-like domain. *J. Dynam. Diff. Equ.* **4**:4, 555–584.

[4] A. Babin and M.Vishik (1990). Attractors of partial differential equations in an unbounded domain. *Proc. Roy. Soc. Edinb. Sec. A* **116**:3-4, 221–243.

[5] A.V. Babin (1995). The attractor of a generalized semigroup generated by an elliptic equation in a tube domain. *Izv. Math.* **44**, 207–223. Translated from Russian (1994) *Izv. Ross. Akad. Nauk Ser. Mat.* **58**, 3–18.

[6] A.V. Babin (2003). Attractors of Navier-Stokes equations. In *Handbook of mathematical fluid dynamics, Vol. II*, S. Friedlander and D. Serre (eds.). North-Holland.

[7] A.V. Babin and M.I. Vishik (1983). Attractors of evolution partial differential equations and estimates of their dimension. *Uspekhi Mat. Nauk* **38**:4(232), 133–187.

[8] A.V. Babin and M.I. Vishik (1992). *Attractors of evolution equations.* Studies in Mathematics and its Applications, 25. North-Holland. Translated and revised from the 1989 Russian original by Babin.

[9] J.M. Ball (1978). On the asymptotic behavior of generalized processes, with applications to nonlinear evolution equations. *J. Differ. Equ.* **27**, 224–265.

[10] J.M. Ball (2004). Global attractors for damped semilinear wave equations. *Discrete Contin. Dyn. Syst.* **10**, 31–52.

[11] J.E. Billotti and J.P. LaSalle (1971). Dissipative periodic processes. *Bull. Amer. Math. Soc.* **77**, 1082–1088.

[12] M. Blair, H. Smith and C. Sogge (2009). Strichartz estimates for the wave equation on manifolds with boundary. *Ann. Inst. Henri Poincaré C* **26**:5, 1817–1829.

[13] M. Blinchevskaya and Yu. Ilyashenko (1999). Estimates for the entropy dimension of the maximal attractor for k-contracting systems in an infinite-dimensional space. *Russ. J. Math. Phys.* **6**, 20–26.

[14] T. Buckmaster and V. Vicol (2019). Non-uniqueness of weak solutions to the Navier-Stokes equation. *Ann. Math.* **189**, 101–144.

[15] N. Burq, G. Lebeau and F. Planchon (2008). Global existence for energy critical waves in 3D domains. *J. Am. Math. Soc.* **21**:3, 831–845.

[16] A.N. Carvalho, J.A. Langa and J.C. Robinson (2013). *Attractors for infinite-dimensional non-autonomous dynamical systems.* Applied Mathematical Sciences. Springer.

[17] P. Brunovsky and P. Polácik (1997). The Morse–Smale structure of a generic reaction-diffusion equation in higher space dimension. *J. Differ. Equ.* **135**, 129–181.

[18] P. Brunovsky and G. Raugel (2003). Genericity of the Morse–Smale property for damped wave equations. *J. Dynam. Differ. Equ.* **15**, 571–658.

[19] V.V. Chepyzhov and M.I. Vishik (1995). Trajectory attractors for evolution equations. *C. R. Acad. Sci. Paris Ser. 1* **321**, 1309–1314.

[20] V. Chepyzhov and M. Vishik (1993). A Hausdorff dimension estimate for kernel sections of non-autonomous evolution equations. *Indiana Univ. Math. J.* **42**:3, 1057–1076.

[21] V.V. Chepyzhov and M.I. Vishik (2002). *Attractors for equations of mathematical physics.* American Mathematical Society Colloquium Publications, 49. American Mathematical Society.

[22] V.V. Chepyzhov and I.A. Ilyin (2004). On the fractal dimension of invariant sets: applications to Navier-Stokes equations. *Discrete Contin. Dyn. Syst.* **10**, 117–135.

[23] I.D. Chueshov (1998). Theory of functionals that uniquely determine long-time dynamics of infinite-dimensional dissipative systems. *Uspekhi Mat. Nauk* **53**, 731–776.

[24] I.D. Chueshov (2015). *Dynamics of quasi-stable dissipative systems.* Universitext, Springer..

[25] I.D. Chueshov and I. Lasiecka (2008). Long-time behavior of second order evolution equations with nonlinear damping. *Mem. Amer. Math. Soc.* **195**(912).

[26] B. Cockburn, D.A. Jones and E.S. Titi (1997). Estimating the number of asymptotic degrees of freedom for nonlinear dissipative systems. *Math. Comp.* **97**, 1073–1087.

[27] P. Constantin and C. Foias (1985). Global Lyapunov exponents, Kaplan–Yorke formulas and the dimension of the attractors for the 2D Navier–Stokes equations. *Commun. Pure Appl. Math.* **38**, 1–27.

[28] H. Crauel and F. Flandoli (1994). Attractors for random dynamical systems. *Probab. Th. Rel. Fields* **100**, 365–393.

[29] A. Douady and J. Oesterlé (1980). Dimension de Hausdorff des attracteurs. *C. R. Acad. Sci. Ser. A-B* **290**:24, A1135–A1138.

[30] A. Eden, C. Foias, B. Nicolaenko and R. Temam (1994). *Exponential attractors for dissipative evolution equations.* Wiley.

[31] M. Efendiev, A. Miranville and S. Zelik (2004). Global and exponential attractors for nonlinear reaction-diffusion systems in unbounded domains. *Proc. Roy. Soc. Edinb. Sec. A* **134**:2, 271–315.

[32] E. Feireisl (1994). Attractors for semilinear damped wave equations on R3. *Nonlinear Anal.* **23**:2, 187–195.

[33] B. Fiedler and C. Rocha (2008). Connectivity and design of planar global attractors of Sturm type. II: connection graphs. *J. Differ. Equ.* **244**, 1255–1286.

[34] C. Foia and G. Prodi (1967). Sur le comportement global des solutions non-stationnaires des quations de Navier-Stokes en dimension 2. *Rend. Sem. Mat. Univ. Padova* **39**, 1–34.

[35] C. Foias and R. Temam (1979). Some analytic and geometric properties of the solutions of the evolution Navier-Stokes equations. *J. Math. Pure et Appl.* **58**, 339–368.

[36] C. Foias and R. Temam (1984). Determination of the solutions of the Navier-Stokes equations by a set of nodal values. *Math. Comp.* **43**(167), 117–133.

[37] C. Foias, G. Sell and R. Temam (1988). Inertial manifolds for nonlinear evolutionary equations. *J. Differ. Equ.* **73**:2 309–353.

[38] C. Foias and E.S. Titi (1991). Determining nodes, finite difference schemes and inertial manifolds. *Nonlinearity* **4**:1, 135–153.

[39] C. Foias, O.P. Manley, R. Rosa and R. Temam (2001). *Navier-Stokes Equations and Turbulence.* Cambridge University Press.

[40] R.L. Frank, D. Hundertmark, M. Jex and P. Nam (2021). The Lieb–Thirring inequality revisited. *J. Eur. Math. Soc.* **23**, 2583–2600.

[41] J. Ghidaglia and R. Temam (1987). Attractors for damped nonlinear hyperbolic equations. *J. Math. Pures Appl.* **66**, 273–319.

[42] J. Hale (1977). *Theory of functional differential equations*, Second Edition. Applied Mathematical Sciences. Springer-Verlag.

[43] J. Hale (1985). Asymptotic behaviour and dynamics in infinite dimensions. In *Nonlinear Differential Equations*, J.K. Hale and P. Martinez-Amores (eds.). Research Notes in Mathematics. Pitman.

[44] J. Hale (1988). *Asymptotic behavior of dissipative systems.* Mathematical Surveys and Monographs, 25. American Mathematical Society.

[45] J. Hale (1997). Dynamics of a scalar parabolic equation. Proceedings of the Geoffrey J. Butler Memorial Conference in Differential Equations and Mathematical Biology (Edmonton, AB, 1996). *Canad. Appl. Math. Quart.* **5**:3, 209–305.

[46] A. Haraux (1985). Two remarks on dissipative hyperbolic problems. In *Nonlinear Partial Differential Equations and Their Applications*, H. Brezis and J.L. Lions (eds.). College de France Seminar, vol. 7. Pitman.

[47] A. Haraux (1988). Attractors of asymptotically compact processes and applications to nonlinear partial differential equations. *Commun. Partial. Differ. Equ.* **13**:11, 1383–1414.

[48] A. Haraux (1991). Systèmes dynamiques dissipatifs et applications. Research in Applied Mathematics, 17. Masson.

[49] B. Hunt (1996). Maximum local Lyapunov dimension bounds the box dimension of chaotic attractors. *Nonlinearity* **9**, 845–852.

[50] M.G. Grillakis (1990). Regularity and asymptotic behaviour of the wave equation with a critical nonlinearity. *Ann. Math.* **132**:3. 485–509.

[51] Yu. S. Il'yashenko (1992). Weakly contracting systems and attractors of Galerkin approximations of Navier-Stokes equations on the two-dimensional torus. *Selecta Math. Soviet* **11**:3, 203–239. Translation from (1982) *Usp. Mekh.* **5**:1, 31–63.

[52] Yu. Ilyashenko (1983). On the dimension of attractors of k-contracting systems in an infinite-dimensional space (Russian). *Vestn. Mosk. Univ., Ser. I* **3**, 52–59.

[53] M.A. Jendoubi (1998). A simple unified approach to some convergence theorems of L. Simon, *J. Funct. Anal.* **153**, 187–202.

[54] H. Jia and V. Sverak (2015). Are the incompressible 3d Navier-Stokes equations locally ill-posed in the natural energy space? *J. Func. Anal.* **268**:12, 3734–3766.

[55] K. Jörgens (1961). Das Anfangswertproblem in Großen für eine Klasse nichtlinearer Wellengleichungen. *Math. Z.* **77**, 295–308.

[56] V. Kalantarov, A. Savostianov and S. Zelik (2016). Attractors for damped quintic wave equations in bounded domains. *Ann. Henri Poincaré* **17**:9, 2555–2584.

[57] A. Kostianko, X. Li, C. Sun and Sergey Zelik (2022). Inertial manifolds via spatial averaging revisited. *SIAM J. Math. Anal.* **54**:1, 268–305.

[58] L.V. Kapitanski (1992). The Cauchy problem for the semilinear wave equation II. *J. Soviet Math.* **62**:3, 2746–2777. Translated from Russian (1990) Zap. Nauchn. Sem. Leningrad. Otdel. Mat. Inst. Steklov. (LOMI) **182**.

[59] L. Kapitanski (1995). Minimal compact global attractor for a damped semilinear wave equation. *Commun. Partial. Differ. Equ.* **20**:7-8, 1303–1323.

[60] P.E. Kloeden and M. Rasmussen (2011). *Nonautomomous dynamical systems.* Mathematical Surveys and Monographs, vol. 176. American Mathematical Society.

[61] O.A. Ladyzhenskaya (1953). *Mixed problems for hyperbolic equation.* (Russian) Nauka.

[62] O.A. Ladyzhenskaya (1959). Solution "in the large" to the boundary-value problem for the Navier-Stokes equations in two space variables. *Commun. Pure Appl. Math.* **12**:3, 427–433. Translated from Russian (1958) *Dokl. Akad. SSSR* **123**:3, 427–429.

[63] O.A. Ladyzhenskaya (1963). *The mathematical theory of viscous incompressible flow.* Gordon and Breach Science Publishers..

[64] O.A. Ladyzhenskaya (1969). An example of non-uniqueness in the class of weak Hopfs solutions for the Navier–Stokes equations. *Izv. Akad. Nauk SSSR, Ser. Matem.* **33**, 240–247.

[65] O.A. Ladyzhenskaya (1972). On a dynamical system generated by Navier–Stokes equations. *Zap. Nauchn. Sem. LOMI* **27**, 91–115.

[66] O.A. Ladyzhenskaja (1983). Limit states for modified Navier-Stokes equations in three-dimensional space. *J. Soviet Math.* **21**, 345–356. Translated from Russian (1979) *Zap. Nauchn. Sem. LOMI* **84**, 131–146.

[67] O.A. Ladyzhenskaya (1985). The finite-dimensionality of bounded invariant sets for the Navier-Stokes system and other dissipative systems. *J. Soviet Math.* **28**, 714–726. Translated from Russian (1982) *Zap. Nauchn. Sem. LOMI* **115**, 137–155.

[68] O.A. Ladyzhenskaya (1987). On the determination of global attractors for the Navier-Stokes equations and other partial differential equations. *Uspekhi Mat. Nauk* **42**:6, 27–73. Translated from Russian (1987) *Uspekhi Mat. Nauk* **42**:6(258), 25–60.

[69] O.A. Ladyzhenskaya (1986). Some directions in the research carried out at the Laboratory of Mathematical Physics of the Leningrad Branch of the Institute of Mathematics (Russian). *Trudy Mat. Inst. Steklov.* **175**, 217–245.

[70] O.A. Ladyzhenskaya (1988). Attractors of nonlinear evolution problems with dissipation. *J. Soviet Math.* **40**:5, 632–640. Translated from Russian (1986) *Zap. Nauchn. Sem. LOMI* **152**, 72–85.

[71] O.A. Ladyzhenskaya (1994). Attractors for the modifications of the three-dimensional Navier-Stokes equations. *Philos. Trans. Roy. Soc. Lond. Ser. A* **346**(1679), 173–190.

[72] O.A. Ladyzhenskaya and G. Sergin (1995). On semigroups generated by initial-boundary value problems describing two-dimensional viscoplastic flows in: Nonlinear evolution equations. *Amer. Math. Soc. Trans. Ser. 2* **164**, 99–123.

[73] O.A. Ladyzhenskaya and G.A. Seregin (1998). Smoothness of solutions of equations describing generalized Newtonian flows and estimates for the dimensions of their attractors. *Izv. Math.* **62**:1, 55–113.

[74] J. Leray (1934). Sur le mouvement d'un liquide visqueux emplissant l'espace. *Acta Math.* **63**, 193–248.

[75] E. Lieb (1984). On characteristic exponents in turbulence. *Commun. Math. Phys.* **92**, 473–480.

[76] J. L. Lions and W. A. Strauss (1965). Some non-linear evolution equations. *Bull. Soc. Math. Fr.* **93**, 43–96.

[77] V.X. Liu (1993). A sharp lower bound for the Hausdorff dimension of the global attractors of the 2D Navier–Stokes equations. *Commun. Math. Phys.* **158**, 327–339.

[78] E. Lorenz (1963). Deterministic nonperiodic flow. *J. Atmos. Sci.* **20**:2, 130–141.

[79] J. Mallet-Paret (1976). Negatively invariant sets of compact maps and an extension of a theorem of Cartwright. *J. Differ. Equ.* **22**:2, 331–348.

[80] J. Malek and J. Necas (1996). A finite-dimensional attractor for three-dimensional flow of incompressible fluids. *J. Differ. Equ.* **127**:2, 498–518.

[81] V.S. Melnik and J. Valero (1998). On attractors of multivalued semi-flows and differential inclusions. *Set-Valued Anal.* **6**, 83–111.

[82] A. Miranville and S. Zelik (2008). Attractors for dissipative partial differential equations in bounded and unbounded domains. In *Handbook of differential equations, Vol. IV*, F. Battelli and M. Fečkan (eds.). Elsevier/North-Holland.

[83] A. Mielke and G. Schneider (1995). Attractors for modulation equations on unbounded domains-existence and comparison. *Nonlinearity* **8**, 743–768.

[84] I. Moise, R. Rosa and X. Wang (1998). Attractors for non-compact semigroups via energy equations. *Nonlinearity* **11**:5, 1369–1393.

[85] J. Robinson (2001). *Infinite-dimensional dynamical systems: An introduction to dissipative parabolic PDEs and the theory of global attractors.* Cambridge Texts in Applied Mathematics. Cambridge University Press.

[86] J. C. Robinson, J. L. Rodrigo and W. Sadowski (2016). *The three-dimensional Navier-Stokes equations, Classical theory.* Cambridge Studies in Advanced Mathematics. Cambridge University Press.

[87] R. Rosa (1998). The global attractor for the 2D Navier-Stokes flow on some unbounded domains. *Nonlinear Anal.* **32**, 71–85.

[88] J. Sather (1966). The existence of a global classical solution of the initial-boundary value problem for $\Box u + u^3 = f$. *Arch. Rational Mech. Anal.* **22**, 292–307.
[89] G. Sell (1996). Global attractors for the three-dimensional Navier-Stokes equations. *J. Dynam. Differ. Equ.* **8**, 1–33.
[90] T. Tao (2016). Finite-time blowup for a supercritical defocusing nonlinear wave system. *Anal. PDE* **9**:8, 1999–2030.
[91] R. Temam (1997). *Infinite-dimensional dynamical systems in mechanics and physics*, second edition. Applied Mathematical Sciences, 68. Springer-Verlag.
[92] M. Vishik, S. Zelik and V. Chepyzhov (2013). Regular attractors and non-autonomous perturbations of them. *Sb. Math.* **204**:1, 1–42.
[93] G. F. Webb (1979). A bifurcation problem for a nonlinear hyperbolic partial differential equation. *SIAM J. Math. Anal.* **10**:5, 922–932.
[94] S. Zelik (2004). Asymptotic regularity of solutions of a nonautonomous damped wave equation with a critical growth exponent. *Commun. Pure Appl. Anal.* **3**:4, 921–934.
[95] S. Zelik (2004). Asymptotic regularity of solutions of singularly perturbed damped wave equations with supercritical nonlinearities. *Discrete Cont. Dyn. Sys.* **11**:2-3, 351–392.
[96] S. Zelik (2013). Infinite energy solutions for damped Navier-Stokes equations in $\mathbb{R}^2$. *J. Math. Fluid Mech.* **15**:4, 717–745.
[97] S. Zelik (2014). Inertial manifolds and finite dimensional reduction for dissipative PDEs. *Proc. Roy. Soc. Edinb. Sec. A* **144**:6, 1245–1327.

Preface

These lecture notes are devoted to questions of the behaviour, when $t \to \infty$, of trajectories $V_t(v), t \in \mathbb{R}^+ = [0, \infty)$ for semigroups $\{V_t,\ t \in \mathbb{R}^+,\ X\}$ of non-linear bounded continuous operators V_t in a locally non-compact metric space X and for solutions of abstract evolution equations. The latter contain initial boundary value problems for many dissipative evolutionary PDE (partial differential equations).

In contrast to the traditional theory of the local stability of PDE (i.e. in the vicinity of a solution) we study the behaviour of all trajectories or solutions for the problems and give a description of the set of all limit states. We will not make assumptions either about the smallness of the parameters in the problem or on the closeness of the problem to a linear one, neither will we consider any other condition that ensures that all the solutions of the problem tend to some special solution. Our purpose is to develop a global theory of stability for dissipative problems of mathematical physics. The principal ideas in this subject were formulated in paper [1] and I follow them here. The object of paper [1] concerns boundary value problems for Navier–Stokes equations. This object helped us to understand which properties of semigroup $\{V_t,\ t \in \mathbb{R}^+,\ X\}$ imply the compactness of the set of all limit states (or, which is the same, the minimal global B-attractor), its invariance, the possibility of continuing the semigroup restricted on $\mathcal{M}$ to the full group on $\mathcal{M}$ and a finiteness of dynamics $\{V_t,\ t \in \mathbb{R} = (-\infty, +\infty),\ \mathcal{M}\}$.

The latter was a source of investigations of the finiteness of dimensions of compact sets in the Hilbert space X which are invariant under a nonlinear bounded operator V enjoying some special properties. The paper by Mallet-Paret [2] was the first one in this direction. In the paper by Douady and Oesterlé [3], the Hausdorff dimension ($\dim_H \mathcal{A}$) of such sets $\mathcal{A}$ was estimated for a wider class of operators V. Although the full proof in [3] was done for the case of a Euclidean space $X \equiv \mathbb{R}^n$, the authors pointed out that it may be generalized to the case of a Hilbert space X.

After these papers many works devoted to such question were published ([4]–[10], etc.). In the first part of Chapter 4 we evaluate Hausdorff and fractal dimensions of compact invariant sets following the approach of paper [3] (see [11]). In the second part of Chapter 4 we show how to verify the conditions of our theorems in the case of semigroups generated by evolution equations.

Let us mention that in paper [6] there is a theorem with a very short and clear proof which was used for estimating both $\dim_H \mathcal{A}$ and the fractal dimension for many PDE of different types. But majorants obtained through this theorem are worse than majorants deduced from theorems of Chapter 4 (see Part II of these lectures and [11]).

We do not give here the full list of papers relating to attractors of PDE. In the eighties several papers on this subject have been published, and the number continually increases. A survey of the relevant papers published before 1986 can be found in [12]. I would like to point out that there are many connections between the results of Chapters 2 and 3 and the results of American mathematicians from Brown University. The latter were developed in the process of studying ODE (ordinary differential equations) with delay and abstract discrete semigroups. They have been expounded in the monograph [13] by J.K. Hale. Semi-linear parabolic equations (mostly with one space argument) are considered in the book [14] by D. Henry, who is concerned only with the investigations of American mathematicians and does not seem to be aware of paper [1]; actually in the first lines of [14] he expresses the wish that attractors of the Navier–Stokes equations and some other problems of hydrodynamics for viscous fluids be investigated

I would like to express my cordial thanks to members of the Academia Nazionale dei Lincei and to Professor G. Fichera especially for the invitation to deliver these lectures and to publish them. I am very much obliged to Professors G. Fichera, P. Castellani and M. Sneider for their help in the preparation of the English version of my lectures.

PART I

Attractors for the semigroups of operators: an abstract framework

1

Basic notions

In this chapter we shall deal with semigroups $\{V_t, t \in \mathbb{R}^+ = [0, +\infty)\}$ of continuous operators $V_t : X \to X$ acting on a complete metric space X. We shall denote them $\{V_t, t \in \mathbb{R}^+, X\}$ or simply $\{V_t\}$.

In what follows, the term *semigroup* refers to any family of continuous operators $V_t : X \to X$ depending on a parameter $t \in \mathbb{R}^+$ and enjoying the semigroup property: $V_{t_1}(V_{t_2}(x)) = V_{t_1+t_2}(x)$ for all $t_1, t_2 \in \mathbb{R}^+$ and $x \in X$.

A semigroup $\{V_t\}$ is called *pointwise continuous* if the mapping $t \to V_t(x)$ from $\mathbb{R}^+$ to X is continuous for each $x \in X$. A semigroup is called *continuous* if the mapping $(t, x) \to V_t(x)$ from $\mathbb{R}^+ \times X$ to X is continuous.

Given a semigroup $\{V_t\}$ the following notation will be frequently used:

$$\begin{aligned}
\gamma^+(x) &:= \{y \in X \mid y = V_t(x), t \in \mathbb{R}^+\} \equiv \{V_t(x), t \in \mathbb{R}^+\};\\
\gamma^+_{[t_1, t_2]}(x) &:= \{V_t(x), t \in [t_1, t_2]\};\\
\gamma^+_t(x) &:= \gamma^+_{[t, \infty)}(x) \equiv \{V_\tau(x), \tau \in [t, \infty)\};\\
\gamma^+(A) &:= \bigcup_{x\in A} \gamma^+(x);\\
\gamma^+_{[t_1, t_2]}(A) &:= \bigcup_{x\in A} \gamma^+_{[t_1, t_2]}(x);\\
\gamma^+_t(A) &:= \bigcup_{x\in A} \gamma^+_t(x).
\end{aligned}$$

It is easy to verify that $V_t(\gamma^+(A)) = \gamma^+_t(A)$.
The curve $\gamma^+(x)$ is called the positive semi-trajectory of x.
The collection of all bounded subsets of X is denoted by $\mathcal{B}$.

We use the letter B (with or without indices) to denote the elements of $\mathcal{B}$, i.e. the bounded subsets of X.

A semigroup $\{V_t\}$ is called *locally bounded* if $\gamma^+_{[0,\,t]}(B) \in \mathcal{B}$ for all $B \in \mathcal{B}$ and all $t \in \mathbb{R}^+$. $\{V_t\}$ is a *bounded semigroup* if $\gamma^+(B) \in \mathcal{B}$ for each $B \in \mathcal{B}$.

Let A and M be subsets of X. We say that *A attracts M* or *M is attracted* to A by semigroup $\{V_t\}$ if for every $\epsilon > 0$ there exists a $t_1(\epsilon, M) \in \mathbb{R}^+$ such that $V_t(M) \subset \mathcal{O}_\epsilon(A)$ for all $t \geq t_1(\epsilon, M)$. Here $\mathcal{O}_\epsilon(A)$ is the ϵ-neighbourhood of A (i.e. the union of all open balls of radii ϵ centered at the points of A). We say that the set $A \subset X$ *attracts the point* $x \in X$ if A attracts the one-point set $\{x\}$.

If A attracts each point x of X then A is called a *global attractor* (for the semigroup). A is called a *global B-attractor* if A attracts each bounded set $B \in \mathcal{B}$.

A semigroup is called *pointwise dissipative* (respectively, *B-dissipative*) if it has a bounded global attractor (respectively a bounded global B-attractor).

Our main purpose here is to find those semigroups for which there is a *minimal closed global B-attractor* and investigate properties of such attractors. These attractors will be denoted by $\mathcal{M}$. We shall examine also the existence of a *minimal closed global attractor* $\widehat{\mathcal{M}}$. It is clear that $\widehat{\mathcal{M}} \subset \mathcal{M}$ and later on we will also verify that $\widehat{\mathcal{M}}$ might be just a small part of $\mathcal{M}$.

The concept of invariant sets is closely related to these subjects. We call a set $A \subset X$ *invariant* (relative to semigroup $\{V_t\}$) if $V_t(A) = A$ for all $t \in \mathbb{R}^+$.

A set $A \subset X$ is called *absorbing* if for every $x \in X$ there exists a $t_1(x) \in \mathbb{R}^+$ such that $V_t(x) \in A$ for all $t \geq t_1(x)$. A set A is called *B-absorbing* if for every $B \in \mathcal{B}$ there exists a $t_1(B) \in \mathbb{R}^+$ such that $V_t(B) \subset A$ for all $t \geq t_1(B)$.

In our investigation of the problems concerning the attractors $\mathcal{M}$ and $\widehat{\mathcal{M}}$ the concept of ω-limit sets will play a fundamental role. For $x \in X$ the *ω-limit set* $\omega(x)$ is, by definition, the set of all $y \in X$ such that $y = \lim_{k\to\infty} V_{t_k}(x)$ for a sequence $t_k \nearrow +\infty$.

The *ω-limit set* $\omega(A)$ for a set $A \subset X$ is the set of the limits of all converging sequences of the form $V_{t_k}(x_k)$, where $x_K \in A$ and $t_k \nearrow +\infty$.

An equivalent description of the ω-limit sets is given by

Lemma 1.1

$$\omega(x) = \bigcap_{t\geq 0} [\gamma_t^+(x)]_X; \quad \omega(A) = \bigcap_{t\geq 0} [\gamma_t^+(A)]_X. \tag{1.1}$$

Here the symbol $[\ ]_X$ means the closure in the topology of the metric space X.

The proof of Lemma 1.1 is traditional and so is omitted. Since, $\gamma_{t_2}^+(A) \subset \gamma_{t_1}^+(A)$ whenever $t_2 > t_1$, the intersection over all $t \in \mathbb{R}^+$ in (1.1) may be replaced by $\bigcap_{t\geq T}$ with any $T \in \mathbb{R}^+$.

It is necessary to have in mind that for locally non-compact spaces X the use of the concept of limit sets requires some caution since the intersection $A_0 = \bigcap_{k=1}^{\infty} A_k$ of $A_k = [A_k]_X \supset A_{k+1} = [A_{k+1}]_X$ in them may be empty (and therefore unhelpful).

2

Semigroups of class $\mathcal{K}$

By definition a semigroup $\{V_t\}$ belongs to the *class* $\mathcal{K}$ if for each $t > 0$ the operator V_t is compact, i.e. for any bounded set $B \subset X$ its image $V_t(B)$ is precompact. (I should remind readers that the operators V_t are supposed to be continuous, see the beginning of Chapter 1).

Theorem 2.1 *Let the semigroup $\{V_t\}$ belong to the class $\mathcal{K}$. Let $A \subset X$ and $T \in \mathbb{R}^+$. Suppose that $\overset{+}{\gamma_T}(A) \in \mathcal{B}$. Then*

(i) $\omega(A)$ is non-empty and compact,
(ii) $\omega(A)$ attracts A,
(iii) $\omega(A)$ is invariant, i.e. $V_t(\omega(A)) = \omega(A)$ for all $t \in \mathbb{R}^+$,
(iv) $\omega(A)$ is the minimal closed set which attracts A,
(v) $\omega(A)$ is connected provided A is connected and the semigroup $\{V_t\}$ is continuous.

Proof The sets $V_t(\overset{+}{\gamma_T}(A)) = \overset{+}{\gamma}_{t+T}(A)$, $0 < t < +\infty$, are precompact and $\overset{+}{\gamma}_{t_2+T}(A) \subset \overset{+}{\gamma}_{t_1+T}(A)$ for all $t_2 > T_1$. Therefore $\omega(A) = \bigcap_{t>0}[\overset{+}{\gamma}_{t+T}(A)]_X$ is the intersection of an ordered family of compact sets. Hence $\omega(A)$ is non-empty, compact and attracts A.

In order to prove (iii) we verify immediately that $V_t(\omega(A)) \subset \omega(A)$. Actually, if $y \in \omega(A)$ then $y = \lim_{k\to\infty} V_{t_k}(x_k)$ for some $x_k \in A$ and $t_k \nearrow +\infty$, hence $V_t(y) = V_t(\lim_{k\to\infty} V_{t_k}(x_k)) = \lim_{k\to\infty} V_{t+t_k}(x_k)$ and thus $V_t(y) \in \omega(A)$.

To obtain the reverse inclusion, $\omega(A) \subset V_t(\omega(A))$, let $x \in \omega(A)$ and therefore $x = \lim_{k\to\infty} V_{t_k}(x_k)$ for some $x_k \in A$ and $t_k \nearrow +\infty$. We may assume that $1 + T + t \leq t_1 < t_2 \ldots$. The points $y_k = v_{t_k - t}(x_k)$, $k = 1, 2\ldots$, belong to the precompact set $\overset{+}{\gamma}_{T+1}(A)$. Hence there is a converging subsequence $\{y_{k_j}\}$ and $\lim_{j\to\infty} y_{k_j} = y \in \omega(A)$.

Consequently $x = \lim_{j\to\infty} V_{t_{k_j}}(x_{k_j}) = \lim_{j\to\infty} V_t(y_{k_j}) = V_t(y)$. Thus we obtain $\omega(A) \subset V_t(\omega(A))$ and (iii) is proved.

To prove the minimality of $\omega(A)$ suppose the contrary and let F be a proper closed subset of $\omega(A)$ which attracts A. As $\omega(A)$ is compact so is F. Choose any $y \in \omega(A)\backslash F$. For $\epsilon > 0$ small enough the ϵ-neighbourhoods $\mathcal{O}_\epsilon(y)$ and $\mathcal{O}_\epsilon(F)$ do not intersect. We assumed that F attracts A. Hence $V_t(A) \subset \mathcal{O}_\epsilon(F)$ for all $t \geq t(\epsilon)$ with some $t(\epsilon) \geq 0$. On the other hand $y = \lim_{k\to\infty} V_{t_k}(x_k)$ for some $x_k \in A$ and $t_k \nearrow +\infty$ (since $y \in \omega(A)$). Consequently, $V_{t_k}(A) \cap \mathcal{O}_\epsilon(y) \neq \varnothing$ for t_k large enough. Hence $\mathcal{O}_\epsilon(F) \cap \mathcal{O}_\epsilon(y) \neq \varnothing$, a contradiction.

Now let A be connected and the semigroup be continuous. Suppose that $\omega(A)$ is not connected. Then we may decompose $\omega(A)$ as follows: $\omega(A) = F_1 \cup F_2$, where F_1 and F_2 are non-empty closed disjoint sets. Therefore open ϵ-neighbourhoods $\mathcal{O}_\epsilon(F_1)$ and $\mathcal{O}_\epsilon(F_2)$ do not intersect for $\epsilon > 0$ small enough. We have $\mathcal{O}_\epsilon(\omega(A)) = \mathcal{O}_\epsilon(F_1) \cup \mathcal{O}_\epsilon(F_2)$.

Since $\omega(A)$ attracts A, there is some $t_1 \equiv t_1(\epsilon, A)$ such that $\gamma_t^+(A) \subset \mathcal{O}_\epsilon(\omega(A))$ for all $t \geq t_1$. But $\gamma_t^+(A)$ is connected since it is the continuous image of $[t, +\infty) \times A$ (under the mapping $(\tau, x) \to V_\tau(x)$). So for all $t \geq t_1$ either $\gamma_t^+(A) \subset \mathcal{O}_\epsilon(F_1)$ or $\gamma_t^+(A) \subset \mathcal{O}_\epsilon(F_2)$. Consequently, either $\omega(A) \subset F_1$ or $\omega(A) \subset F_2$, hence either $F_1 = \varnothing$ or $F_2 = \varnothing$; this is a contradiction. Thus, $\omega(A)$ must be connected. □

By definition, a *complete trajectory* $\gamma(x)$ *of the point* x is the curve $x(t)$, $-\infty < t < +\infty$, satisfying the following conditions: $x(t) \in X$ for all $t \in \mathbb{R}$, $x(0) = x$, $V_\tau(x(t)) = x(t+\tau)$ for all $t \in \mathbb{R}$ and $\tau \in \mathbb{R}^+$. The set $\gamma^-(x) = \{x(t), -\infty < t \leq 0\}$ is called a *negative semi-trajectory of* x. Thus, $\gamma(x) = \gamma^+(x) \cup \gamma^-(x)$. In general, for an arbitrary $x \in X$, a complete trajectory $\gamma(x)$ may not exist and even if it does, it might be not unique. However the following statement is true.

Lemma 2.1 *Let A be an invariant set (i.e. $V_t(A) = A$ for all $t \in \mathbb{R}^+$). Then for every $x \in A$ there exists a complete trajectory $\gamma(x)$. If the semigroup $\{V_t\}$ is pointwise continuous then the trajectory $\gamma(x)$ is a continuous curve in A. If the operators V_t, $t \in \mathbb{R}^+$, are invertible on A then:*

(i) through each $x \in A$ passes a unique trajectory $\gamma(x)$;
(ii) the family $\{V_t, t \in \mathbb{R}, A\}$, where $V_t := V_{-t}^{-1}$ for $t < 0$, has the group property: $V_{t_1+t_2} = V_{t_1} V_{t_2}$ for any $t_1, t_2 \in \mathbb{R}$; if additionally A is compact, then $\{V_t, t \in \mathbb{R}, A\}$ is the group of continuous operators. This group is pointwise continuous or continuous if $\{V_t, t \in \mathbb{R}^+, A\}$ is pointwise continuous or continuous correspondingly.

We omit the proof of the lemma since it is traditional. Let us describe only the construction of $\gamma(x)$, $x \in A$. For $x \in A$ there is at least one point $x_{-1} \in A$ for which $V_1(x_{-1}) = x$; for x_{-1} there is a point $x_{-2} \in A$ for which $V_1(x_{-2}) = x_{-1}$ and so on. Let us join the points x_{-k-1} and x_{-k} by the curve $\{V_t(x_{-k-1}), t \in [0,l]\}$. The collection of all these curves for all $k = 0, 1, \ldots$ forms $\gamma^-(x)$ and $x(t) = V_{t+k+1}(x_{-k-1})$, for $t \in [-k-1, -k]$.

Now we turn to the problem of the existence of the minimal global B-attractor $\mathcal{M}$ for a semigroup $\{V_t\}$ of class $\mathcal{K}$.

Consider first the simplest case when there exists a global B-absorbing bounded set $B_0 \in \mathcal{B}$. Then for every $B \in \mathcal{B}$ there is $T(B) \geq 0$ so that $V_t(B) \subset B_0$ for all $t \geq T(B)$. In particular, $V_t(B_0) \subset B_0$ for all $t \geq T(B_0)$ and consequently $\gamma^+_{T(B_0)}(B_0) \subset B_0$. In view of Theorem 2.1 the set $\omega(B_0)$ is a non-empty compact invariant set. Moreover, $\omega(B_0)$ attracts B_0. Hence, for every $\epsilon > 0$ there exists $t_1(\epsilon) \geq 0$ such that $V_t(B_0 \subset \mathcal{O}_\epsilon(\omega(B_0))$ for all $t \geq t_1(\epsilon)$. Therefore, given any $B \in \mathcal{B}$ we have $V_t(B) \subset \mathcal{O}_\epsilon(\omega(B_0))$ for all $t \geq t_1(\epsilon) + T(B)$. Hence, $\omega(B_0)$ is a closed global B-attractor. It is minimal due to Theorem 2.1 (iv). Thus, $\omega(B_0) = \mathcal{M}$.

Assume now that the semigroup $\{V_t\}$ is B-dissipative and B_1 is its bounded global B-attractor. Then $\mathcal{O}_{\epsilon_1}(B_1)$ (with $\epsilon_1 > 0$) is a global B-absorbing bounded set, and due to the first case $\mathcal{M} = \omega(\mathcal{O}_{\epsilon_1}(\mathcal{B}_1))$.

Consider now a more complicated situation. Suppose that $\{V_t\}$ is a bounded and pointwise dissipative semigroup. In particular, there is a bounded global attractor, say B_2. Choose $\epsilon_2 > 0$, and put $\mathcal{B}_1 := \mathcal{O}_{\epsilon_2}(B_2)$ and $B_0 := \gamma^+(B_1)$. We are going to prove that $\omega(B_1) = \mathcal{M}$ is the minimal global B-attractor.

Indeed, since B_2 is a global attractor, for every point $x \in X$ there is $T(x) \geq 0$ such that $V_{T(x)}(x) \in B_1 = \mathcal{O}_{\epsilon_2}(B_2)$. Since B_1 is an open set and the operator $V_{T(x)}$ is continuous, we have $V_{T(x)}(\mathcal{O}_{\epsilon(x)}(x)) \subset B_1$, for some $\epsilon(x) > 0$. Hence $V_{t+T(x)}(\mathcal{O}_{\epsilon(x)}(x)) \subset V_t(B_1) \subset \gamma^+(B_1) = B_0$ for all $t \geq 0$. Now, by standard arguments, for every compact set K there are $\epsilon(K) > 0$ and $T(K) \geq 0$ such that

$$V_t(\mathcal{O}_{\epsilon(K)}(K)) \subset B_0 \qquad \text{for all} \quad t \geq T(K). \tag{2.1}$$

In view of Theorem 2.1 every bounded set B is attracted to its ω-limit set $\omega(B)$. Hence, $V_t(B) \subset \mathcal{O}_{\epsilon_1}(\omega(B))$ for all $t_1 \geq t_1(\epsilon_1, B)$. Since $\omega(B)$ is compact we may choose $\epsilon_1 = \epsilon(\omega(B))$ and deduce from (2.1) that $V_{t+t_1}(B) \subset B_0$ for all $t \geq T(\omega(B))$ and $t_1 = t_1(\epsilon_1, B)$. Thus, B_0 is a global B-absorbing bounded set.

Hence, as was shown above, $\omega(B_0) = \mathcal{M}$. Note that $V_t(B_0) = \gamma_t^+(B_1)$ by definition of B_1 and B_0, and $\gamma_t^+(B_1) \to \omega(B_1)$ when $t \to \infty$, therefore $\omega(B_0) = \omega(B_1)$. For minimality of $\omega(B_1)$ see Theorem 2.1 (iv).

If there is a connected $B \supset \mathcal{M}$ then $\mathcal{M}$ is connected since $V_t(B)$, $t \in \mathbb{R}^+$ are connected and for any $\epsilon > 0$, $\mathcal{M} = V_t(\mathcal{M}) \subset V_t(B) \subset \mathcal{O}_\epsilon(M)$ for $t \geq t_1(\epsilon, B)$.

Thus we have proved the following

Theorem 2.2 *Let $\{V_t, t \in \mathbb{R}^+, X\}$ be a semigroup of class $\mathcal{K}$. Suppose that it is either B-dissipative or bounded and pointwise dissipative. Then $\{V_t, t \in \mathbb{R}^+, X\}$ has a minimal global B-attractor $\mathcal{M}$, which is compact and invariant. $\mathcal{M}$ is connected provided so is X.*

As a by-product of the above considerations (see (2.1)) we have proved:

Proposition 2.1 *If the semigroup $\{V_t, t \in \mathbb{R}^+, X\}$ is bounded and pointwise dissipative then there is a bounded set B_0 such that for every compact K (2.1) holds with some $\epsilon(K) > 0$ and $T(K) \geq 0$ and $V_t(B_0) \subset B_0$ for all $t \in \mathbb{R}^+$.*

The next proposition provides useful information on the structure of the minimal global B-attractor $\mathcal{M}$.

Proposition 2.2 *Under the assumptions of Theorem 2.2 the minimal global B-attractor $\mathcal{M}$ may be characterized as follows:*

(i) $\mathcal{M} = \bigcup_{B \in \mathcal{B}} \omega(B)$;
(ii) $\mathcal{M} = \bigcup_{K \in \mathcal{K}} \omega(K)$ *where $\mathcal{K}$ is the collection of all compact sets in X;*
(iii) $\mathcal{M}$ is the union of all complete bounded trajectories in X;
(iv) $\mathcal{M}$ is the union of all complete precompact trajectories in X;
(v) $\mathcal{M}$ is the maximal invariant bounded set in X;
(vi) $\widehat{\mathcal{M}} = \left[\bigcup_{x \in X} \omega(x)\right]_X$.

Proof

(i) Every bounded set B is attracted to its ω-lirnit set $\omega(B)$ and to $\mathcal{M}$ and, hence, to $\omega(B) \cap \mathcal{M}$. Since $\omega(B)$ is minimal, $\omega(B)$ must entirely lie in $\mathcal{M}$, i.e. $\omega(B) \subset \mathcal{M}$. Now, $\mathcal{M}$ being invariant and compact, we have $\omega(\mathcal{M}) \equiv \mathcal{M}$. Thus, (i) is proved.

(ii) Since $\mathcal{M}$ and $\omega(B)$ for each $B \in \mathcal{B}$, are invariant compact sets, (ii) follows from (i).

(iii) and (iv) In view of Lemma 2.2, through every point $x \in \mathcal{M}$ passes a complete trajectory $\gamma(x)$. Any such trajectory lies in $\mathcal{M}$ and thus is bounded (and precompact). On the other hand, let $\gamma(x) = \{x(t), t \in \mathbb{R}\}$ be a bounded complete trajectory passing through some point $x = x(0) \in X$. Since $\gamma(x)$ is invariant (and bounded), then $\gamma(x)$ is precompact Hence, $B := [\gamma(x)]_X$ is a compact invariant set. Therefore $\omega(B) \equiv B$ and so $B \subset \mathcal{M}$ (see (i) above).

(v) If B is a bounded invariant set, then $V_t(B) = B$ and therefore $\omega(B) = B$ and $B \subset \mathcal{M}$. On the other hand $\mathcal{M}$ is a bounded invariant set.

(vi) is obvious.

This concludes the proof of the proposition. □

The structure of $\mathcal{M}$ is simpler than in the general case, if for the serni-group $\{V_t, t \in \mathbb{R}^+, X\}$ there is a "good" ("strong") Lyapunov function, i.e. a continuous function $\mathcal{L}\colon X \to \mathbb{R}$ which strongly decreases along each $\gamma^+(x)$: $\mathcal{L}(V_t(x)) \searrow$ when $t \nearrow$ (except, of course, stationary points: $z = V_t(Z)$). Let Z be the set of all stationary points of $\{V_t\}$. If the semigroup $\{V_t\}$ belongs to the class $\mathcal{K}$ and $\gamma^+(x) \in \mathcal{B}$ for any $x \in X$, then Z is its minimal global attractor $\widehat{\mathcal{M}}$. In fact, for any $x \in X$ there exists a $\lim_{t\to\infty} \mathcal{L}(V_t(x)) \equiv \ell_+(x)$, a compact $\omega(x)$ and $\mathcal{L}_{|\omega(x)} \equiv \ell_+(x) = \text{constant}$. Therefore $\omega(x) \subset Z$ and x is attracted to Z. If Z is a bounded set then the semigroup $\{V_t\}$ is pointwise dissipative and Theorem 2.2 guarantees the existence of a compact attractor $\mathcal{M}$ provided $\{V_t\}$ is bounded. For each $x \in \mathcal{M}$, we can take a complete trajectory $\gamma(x) = \{x(t), t \in \mathbb{R}, x(0) = x\}$ lying in $\mathcal{M}$ and determine for it the α-limit set $\alpha(\gamma(x)) := \bigcap_{\tau\leq 0}[\gamma_\tau^-(x)]_X$ where $\gamma_\tau^-(x) := \{x(t), t \leq \tau\}$. This α-limit, set like, $\omega(x)$ is nonempty, invariant and $x(t) \to \alpha(\gamma(x))$ when $t \to -\infty$. It also belongs to Z since $\lim_{t\to-\infty} L(x(t)) = \text{constant} = \mathcal{L}_{|\alpha(\gamma(x))}$. So we may say that both ends of trajectory $\gamma(x)$ tend to Z. If for example, the space X is connected and Z is not connected, then $\mathcal{M}$ (which is connected) contains not only points of Z but complete trajectories connecting points of Z (so Z is smaller than $\mathcal{M}$).

Let us summarize these facts:

Theorem 2.3 *Suppose that the semigroup $\{V_t, t \in \mathbb{R}^+, X\}$ belongs to the class $\mathcal{K}$ and $\gamma^+(x) \in \mathcal{B}$ for any $x \in X$. If for this semigroup there is a "good" Lyapunov function $\mathcal{L}$, then its minimal global attractor $\widehat{\mathcal{M}}$ is non-empty and coincides with the set Z of all stationary points. If Z is a bounded set and $\{V_t\}$ is bounded then the semigroup has a minimal global B-attractor $\mathcal{M}$ enjoying properties indicated in Theorem 2.2. Both ends of any complete trajectory $\gamma(x) \subset \mathcal{M}$ tend to Z (when $t \to \pm\infty$ correspondingly). If X is connected and Z is not, then Z is a proper part of $\mathcal{M}$ and the, attractor $\mathcal{M}$ consists of complete trajectories which connect points of Z.*

3

Semigroups of class $A\mathcal{K}$

A semigroup $\{V_t, t \in \mathbb{R}^+, X\}$ belongs to the class $A\mathcal{K}$ (or it is asymptotically compact) if it possesses the following property: for every $B \in \mathcal{B}$ such that $\gamma^+(B) \in \mathcal{B}$, each sequence of the form $\{V_{t_k}(x_k)\}_{k=1}^{\infty}$, where $x_k \in B$ and $t_k \nearrow +\infty$, is precompact.

Here we restrict ourselves to the case of continuous semigroups of class $A\mathcal{K}$. We begin with two elementary propositions concerning continuous semigroups.

Proposition 3.1 *For every compact set K and $t \in \mathbb{R}^+$ the set $\gamma^+_{[0,\,t]}(K)$ is compact.*

The proof is evident

Proposition 3.2 *If K is compact and $\gamma^+(K)$ is precompact then $\omega(K)$ is a non-empty compact invariant set attracting K.*

Proof In fact, this proposition was proved in Chapter 2. Denoting $K_1 := [\gamma^+(K)]_X$, we know this is compact and $V_t(K_1) \subset K_1$. So we have a semigroup $\{V_t, t \in \mathbb{R}^+, K_1\}$ of continuous operators V_t acting on a metric space K_1. Hence, this semigroup is of class $\mathcal{K}$ and we may apply Theorem 2.1 to obtain the desired result. □

Now we pass to the semigroups of class $A\mathcal{K}$.

Proposition 3.3 *Let $\{V_t, t \in \mathbb{R}^+, X\}$ be a continuous semigroup of class $A\mathcal{K}$. Suppose that K is a compact set such that the $\gamma^+(K)$ is bounded. Then $\gamma^+(K)$ is precompact and thus the statement of Proposition 3.2 is true.*

Proof Let y_n, $n = 1, 2, \ldots$, be an arbitrary sequence of points from $\gamma^+(K)$, i.e. $y_n = V_{t_n}(x_n)$ for some $x_n \in K$ and $t_n \in \mathbb{R}^+$. If the set $\{t_n\}_{n=1}^{\infty}$ is bounded, the set $\{y_n\}_{n=1}^{\infty}$ is precompact by Proposition 3.1. If the set $\{t_n\}_{n=1}^{\infty}$ is unbounded, then we may choose a subsequence $t_{n_j} \nearrow +\infty$ and the set $\{V_{t_{n_j}}(x_{n_j})\}_{j=1}^{\infty}$ will be precompact due to the fact that the semigroup $\{V_t\}$ is of class $A\mathcal{K}$. □

Proposition 3.4 *Let $\{V_t\}$ be a continuous semigroup of class $A\mathcal{K}$. Suppose that $\gamma^+(B) \in \mathcal{B}$ for some $B \in \mathcal{B}$. Then $\omega(B)$ is a non-empty invariant compact set attracting B and $\omega(B)$ is connected if B is connected.*

Proof As $\gamma^+(x) \in \mathcal{B}$ for each $x \in B$, the ω-limit sets $\omega(x)$, $x \in B$, are non-empty, and hence $\omega(B) \neq \varnothing$. It is evident that $\omega(B)$ is closed and bounded. To prove that it is invariant we have only to check the embedding $\omega(B) \subset V_t(\omega(B))$ since the inverse embedding is always valid provided operators V_t are continuous (see the proof of Theorem 2.1). Choose an arbitrary $y \in \omega(B)$. We know that $y = \lim_{n\to\infty} V_{t_n}(x_n)$ for some $x_n \in B$ and some $t_n \nearrow +\infty$. Obviously, $V_{t_n}(x_n) = V_t(V_{t_n - t}(x_n))$ if $t_n \geq t$. The set $\{V_{t_n - t}(x_n)\}_{t_n \geq t}$ is precompact since $\{V_t\}$ is of class $A\mathcal{K}$.

Choose a converging subsequence $\{V_{t_{n_j}-t}(x_{n_j})\}_{j=1}^{\infty}$ and let $z = \lim_{j\to\infty} V_{t_{n_j}-t}(x_{n_j})$. Clearly $z \in \omega(B)$ and $V_t(z) = y$. Thus the embedding $\omega(B) \subset V_t(\omega(B))$ is established. Hence $\omega(B)$ is invariant. From this it follows that each sequence $\{x_k\}_{k=1}^{\infty}$ with $x_k \in \omega(B)$ may be represented as $\{x_k = V_k(\overline{x}_k)\}_{k=1}^{\infty}$ with $\overline{x}_k \in \omega(B)$ and therefore it is precompact, so that $\omega(B)$ is compact.

It remains to prove that $\omega(B)$ attracts B. Suppose that it is not true. Then we can choose a sequence $\{V_{t_k}(x_k)\}_{k=1}^{\infty}$ with $x_k \in B$ and $t_k \nearrow +\infty$ so that dist $\{\{V_{t_k}(x_k)\}_{k=1}^{\infty}; \omega(B)\} \geq \epsilon > 0$ for some ϵ. The asymptotical compactness of our semigroup $\{V_t\}$ implies the precompactness of the set $\{V_{t_k}(x_k)\}$. Since all the limit points of the set $\{V_{t_k}(x_k)\}_{k=1}^{\infty}$ must lie in $\omega(B)$, the distance between $\omega(B)$ and $\{V_{t_k}(x_k)\}_{k=1}^{\infty}$ is zero. This is a contradiction.

If B is connected then $\omega(B)$ is connected by using the same arguments as in the proof of Theorem 2.1 (v). □

Theorem 3.1 *Let $\{V_t, t \in \mathbb{R}^+, X\}$ be a continuous bounded and pointwise dissipative semigroup of class $A\mathcal{K}$. Then there exists a non-empty minimal global B-attractor $\mathcal{M}$. $\mathcal{M}$ is compact and invariant. If X is connected then $\mathcal{M}$ is also connected.*

Proof By Proposition 2.1 there is a bounded set B_0 such that for every compact K

$$V_t(\mathcal{O}_{\epsilon(K)}(K)) \subset B_0 \qquad \text{for all} \quad t \geq T(K), \tag{3.1}$$

with some $\epsilon(K) > 0$ and $T(K) < +\infty$, and, in addition, $V_t(B_0) \subset B_0$ for all $t \in \mathbb{R}^+$.

Since the semigroup $\{V_t\}$ is bounded (i.e. $\gamma^+(B) \in \mathcal{B}$ for any $B \in \mathcal{B}$), Proposition 3.4 yields that for every $B \in \mathcal{B}$ the ω-limit set $\omega(B)$ is non-empty, compact, invariant and attracts B. In particular, so is $\omega(B_0)$. We claim that $\omega(B_0) \equiv \mathcal{M}$. To prove this statement we need only to show that $\omega(B_0)$ attracts each bounded set. But if $B \in \mathcal{B}$ then B is attracted to the compact set $\omega(B) \equiv K$. Hence, $V_t(B) \subset \mathcal{O}_{\epsilon(K)}(K)$ for all $t \geq t_1(B)$. Because of (3.1), $V_{t+t_1(B)}(B) \subset B_0$ for all $t \geq T(K)$. But we know that B_0 is attracted to $\omega(B_0)$. Hence, B is attracted to $\omega(B_0)$ as well.

If X is connected then we may choose a bounded connected set $B_1 \supset B_0$. Its ω-limit set $\omega(B_1)$ is connected and it is easy to verify that $\omega(B_1) \equiv \omega(B_0)$. □

Theorem 3.2 *If $\{V_t, t \in \mathbb{R}^+, X\}$ is a continuous bounded semigroup of class $\mathcal{AK}$ and it has a "good" Lyapunov function $\mathcal{L}\colon X \to \mathbb{R}$, then all statements of Theorem 2.3 are true for it.*

The proof of the theorem is the same as for Theorem 2.3 if we bear in mind the results of Theorem 3.1.

The following theorem is useful for applications:

Theorem 3.3 *Suppose that the semigroup $\{V_t, t \in \mathbb{R}^+, M\}$ is defined on a subset M of a Banach space X with a norm $\|\cdot\|_X$. Suppose also that V_t can be decomposed in the sum $W_t + U_t$, where $\{W_t, t \in \mathbb{R}^+, M\}$ is a family of operators such that for any bounded set $B \subset M$*

$$\|W_t(B)\|_X \leq m_1(t)m_2(\|B\|_X), \tag{3.2}$$

where $m_k\colon \mathbb{R}^+ \to \mathbb{R}^+$ are continuous and $m_1(t) \to 0$ when $t \to +\infty$, $\|B\|_X := \sup_{x\in B}\|x\|_X$. The U_t are such that the set $U_t(B)$ is precompact for each bounded set $B \subset M$. Then $\{V_t, t \in \mathbb{R}^+, M\}$ belongs to the class $\mathcal{AK}$.

Proof Let $\gamma^+(B) \in \mathcal{B}$. We show that each set $B_1 := \{V_{t_k}(x_k)\}_{k=1}^{\infty}$, $t_k \nearrow \infty$, $x_k \in B$, can be covered by a finite ϵ-network where ϵ is any positive number. Let us choose ℓ so large that $m_1(\ell) \leq \epsilon[2m_2(\|B\|_X)]^{-1}$ and decompose B_1 in the sum $B_1' U B_1''$, where $B_1' = \{V_{t_k}(x_k)\}_{k=1}^{k_1}$, $t_k < \ell$, and $B_1'' = \{V_{t_k}(x_k)\}_{k=k_1+1}^{\infty}$, $t_{k_1+1} \geq \ell$. B_1'' is a subset of the set $V_\ell(\gamma^+(B))$ and any element of $V_\ell(\gamma^+(B))$ has the form $W_\ell(x) + U_\ell(x)$, where x is an element of $\gamma^+(B)$. The set $U_\ell(\gamma^+(B))$ may be covered by a finite $\epsilon/2$-network since it is precompact and the norms of the elements of $W_\ell(\gamma^+(B))$ are not larger than $\epsilon/2$. Therefore the set $V_\ell(\gamma^+(B))$ may be covered by a finite ϵ-network. Hence B_1 may be covered by a finite ϵ-network as well. □

Afterword

Class $\mathcal{K}$ had appeared in connection with the study of the set of all limit-states for the Navier–Stokes equations ([1], 1972). To this class belong the families of solution operators for many problems of parabolic type. Class $A\mathcal{K}$ had arisen for PDE later (in 80's) during the study of some problems of hyperbolic and mixed types. Class $\mathcal{K}$ is a part of class $A\mathcal{K}$, but we have devoted to it the separate Chapter 2 for historical and methodological reasons. Besides these arguments, the results of Chapter 3 about semigroups of class $A\mathcal{K}$ do not cover the results of Chapter 2, since in Chapter 3 we consider (in contrast to Chapter 2) only continuous semigroups. This restriction is not very important for the theory of attractors and the principal facts of the theory are true for semigroups $\{V_t, t \in \mathcal{T}^+, X\}$ of class $A\mathcal{K}$ with any additive semigroup $\mathcal{T}^+ \subset \mathbb{R}^+$.

Let us formulate, for example, the theorem which generalizes Theorems 2.2 and 3.1.

Definition The semigroup $\{V_t,\, t \in \mathcal{T}^+,\, X\}$ belongs to the class $A\mathcal{K}$ iff for every $B \in \mathcal{B}$ such that $\gamma^+_{T(B)} \in \mathcal{B}$ for a $T(B) \in \mathcal{T}^+$, each sequence of the form $\{V_{t_k}(x_k)\}_{k=1}^{\infty}$ where $x_k \in B$, $t_k \nearrow \infty$, is precompact.

Theorem 3.4 *Let $\{V_t,\, t \in \mathcal{T}^+,\, X\}$ be a point-wise dissipative semigroup of class $A\mathcal{K}$ and suppose that for each $B \in \mathcal{B}$ there exists a $T(B) \in \mathcal{T}^+$ such that $\gamma^+_{T(B)} \in \mathcal{B}$. Then there exists a non-empty minimal global B-attractor $\mathcal{M}$. It is compact and invariant. If X is connected then $\mathcal{M}$ is also connected.*

It is easy to see that if a semigroup has a compact global B-attractor then it has all properties indicated in the conditions of Theorem 3.4.

The proofs of this and other theorems extending the theorems of Chapters 2 and 3 to the semigroups $\{V_t, t \in \mathcal{T}^+, X\}$ of class $A\mathcal{K}$ are close to proofs given here.

4

On dimensions of compact invariant sets

In this chapter we shall estimate $\dim_H(\mathcal{A})$ and $\dim_f(\mathcal{A})$, i.e. Hausdorff and fractal dimensions of compact invariant sets $\mathcal{A}$, and, as a consequence, of attractors $\mathcal{M}$ provided X is a separable Hilbert space.

Let K be a compact set lying in X and $B_r(x)$ the closed ball of radius r centered at x. We associate with every finite covering $\mathcal{U} = \{B_{r_i}(x_i)\}$ of K (i.e. $K \subset \bigcup_i B_{r_i}(x_i)$) the numbers:

$$r(\mathcal{U}) := \max_i r_i,$$

$$n(\mathcal{U}) = \text{number of elements in } \mathcal{U},$$

$$m_{\beta,\, r(\mathcal{U})} := \sum_{i=1}^{n(\mathcal{U})} r_i^\beta$$

and

$$\nu_\beta(\mathcal{U}) := r(\mathcal{U})^\beta n(\mathcal{U}),$$

where β is a positive number.

We shall use the following known lemma.

Lemma 4.1 *Suppose that for the compact set K there is a sequence $\mathcal{U}_s$, $s = 0, 1, \ldots$ of coverings as described above with $r(\mathcal{U}_s) \to 0$, $m_{\beta,\, r(\mathcal{U}_s)}(\mathcal{U}_s) \to 0$ when $s \to \infty$. Then*

$$\dim_H(K) \leq \beta.$$

If for such a sequence $\nu_\beta(\mathcal{U}_s) \to 0$, then

$$\dim_f(K) \leq \beta.$$

Let H be a separable Hilbert space. We shall use the following notations:

$\mathcal{P}^N$ = an orthogonal projector on an N-dimensional subspace $\mathcal{P}^N H$;
$\mathcal{Q}^N H$ = an orthogonal complement to $\mathcal{P}^N H$ and $\mathcal{Q}^N = I - \mathcal{P}^N$;
B_r = the ball in H of radius r centered at the origin;
$B_r(\mathcal{P}^N)$ and $B_r(\mathcal{Q}^N)$ = the analogous balls in the subspaces $\mathcal{P}^N H$ and $\mathcal{Q}^N H$, respectively;
$\mathcal{E}(\mathcal{P}^N, \alpha)$ = the ellipsoid in $\mathcal{P}^N H$ centered at the origin with semi-axes $\alpha_1 \geq \alpha_2 \geq \cdots \geq \alpha_N$, where $(\alpha_1, \ldots, \alpha_N) = \alpha$;
$\mathcal{E}(\mathcal{P}^N, \alpha) \oplus B_\delta(\mathcal{Q}^N)$ is the set of $v \in H$ such that $v = v_1 + v_2$, where $v_1 \in \mathcal{E}(\mathcal{P}^N, \alpha)$ and $v_2 \in B_\delta(\mathcal{Q}^N)$; $v + B = \{v + u \mid u \in B\}$.

In what follows we shall deal with the projectors $\mathcal{P}^N$ and the numbers $\alpha_1, \alpha_2, \ldots$, which depend on the points v of some subsets in H. In this case the dependence on v will be denoted by $\mathcal{P}^N(v)$, $\alpha(v)$, etc.

Let $\alpha_1 \geq \alpha_2 \geq \cdots$ be an infinite sequence of nonnegative real numbers. Then $\omega_k(\alpha)$ denotes the product of the first k numbers of the sequence α, i.e. $\omega_k(\alpha) = \alpha_1 \ldots \alpha_k$, and $\omega_0(\alpha) := 1$. If the numbers α_k depend on the points $v \in \mathcal{A} \subset H$ we use the notation $\overline{\alpha}_k := \sup_{v \in \mathcal{A}} \alpha_k(v)$ and $\overline{\omega}_k := \sup_{v \in \mathcal{A}} \omega_k(\alpha(v))$. It is obvious that $\alpha_k(v) \leq \omega_k^{1/k}(v)$, $\overline{\alpha}_k \leq \overline{\omega}_k^{1/k}$; $\omega_k^{1/k}(v)$ and $\overline{\omega}_k^{1/k}$ do not increase when k grows.

Theorem 4.1 *Let H be a Hilbert space, V a continuous mapping from H into H, and $\mathcal{A}$ a compact subset of H such that $\mathcal{A} \subset V(\mathcal{A})$. Assume that there are $\alpha_1(v), \ldots \alpha_N(v)$, $\delta(v)$, defined on $\mathcal{A}$ and projectors $\mathcal{P}^N(v)$, $v \in \mathcal{A}$, such that $\alpha_1(v) \geq \cdots \geq \alpha_N(v) \geq \delta(v) > 0$ for all $v \in \mathcal{A}$ and*

$$V((v + B_r) \cap \mathcal{A}) \subset V(v) + r[\mathcal{E}(\mathcal{P}^N(v), \alpha(v)) \oplus B_{\delta(v)}(\mathcal{Q}^N(v))] \qquad (4.1)$$

for all $r \leq r_0$ with some r_0 and $N \geq 1$.

If $\overline{\alpha}_1 = \sup_{v \in \mathcal{A}} \alpha_1(v) < +\infty$ and $\overline{\delta} = \sup_{v \in \mathcal{A}} \delta(v) < 1/2$, then

$$\dim_H(\mathcal{A}) \leq \max\left\{N; N \frac{\ln\left[(\sqrt{N}+1)\overline{\omega}_N^{1/N}(\epsilon\overline{\delta})^{-1}\right]}{\ln\left(1/2\overline{\delta}\sqrt{1+\epsilon^2}\right)}\right\} \equiv d_1, \qquad (4.2)$$

where ϵ is an arbitrary number in $(0, 1]$ satisfying the inequality $2\overline{\delta}\sqrt{1+\epsilon^2} < 1$.

If

$$2\sqrt{2}\left(\sqrt{N}+1\right)\overline{\omega}_N^{1/N} \leq 1, \qquad (4.3)$$

then $\dim_H(\mathcal{A}) \leq N$.

Remark Inequality (4.3) implies $\overline{\delta} \leq \overline{\alpha}_N \leq \overline{\omega}_N^{1/N} < 1/(2\sqrt{2})$ and, hence, in (4.2) we may choose $\epsilon = 1$, and, taking into account (4.3), get $\dim_H(\mathcal{A}) \leq N$.

The next theorem provides an estimate for the fractal dimension of $\mathcal{A}$.

Theorem 4.2 *Under the assumptions of Theorem 4.1 suppose $\overline{\alpha}_1 < +\infty$ and $\overline{\delta} < 1/(2\sqrt{2})$. Then*

$$\dim_f(\mathcal{A}) \leq N \ln\left[(\sqrt{N}+1)\chi_N \overline{\delta}^{-1}\right] \Big/ \ln\left(\frac{1}{2\sqrt{2\overline{\delta}}}\right), \tag{4.4}$$

where

$$\chi_N = \max_{\ell=0,\,1,\ldots,\,N}\left(\overline{\omega}_\ell^{1/N} \overline{\delta}^{-1-\ell/N}\right);\ \overline{\omega}_0 := 1. \tag{4.5}$$

In particular, if

$$2\sqrt{2}(\sqrt{N}+1)\chi_N \leq 1, \tag{4.6}$$

then $\dim_f(\mathcal{A}) \leq N$.

Proof of Theorem 4.1 We construct some "pulverizing coverings" $\mathcal{U}_s$, $s = 0, 1, \ldots$ of the set $\mathcal{A}$ with $\rho_s = r(\mathcal{U}_s) \to 0$, $s \to \infty$, and calculate for them the numbers $m_{\beta,\,\rho_s}(\mathcal{U}_s)$. By choosing β sufficiently large we get that $m_{\beta,\,\rho_s}(\mathcal{U}_s) \to 0$ when $s \to \infty$ and $\dim_H(\mathcal{A}) \leq \beta$.

Let $\mathcal{U}_0$ be a covering of $\mathcal{A}$ by a finite number of balls $v_i + B_{r_i}$ $i = 1, \ldots, n(\mathcal{U}_0)$, with $v_i \in \mathcal{A}$ and $r_i \leq r(\mathcal{U}_0) \equiv \rho_0$. The number $m_{\beta,\,\rho_0}(\mathcal{U}_0) = \sum_{i=1}^{n(\mathcal{U}_0)} r_i^\beta$ corresponds to this covering. The next covering is constructed in the following way: since $\mathcal{A} \subset V(\mathcal{A})$ the collection of sets $V((v_i + B_{r_i}) \cap \mathcal{A})$, $i = 1, \ldots, n(\mathcal{U}_0)$, is a covering of $\mathcal{A}$. Because of the assumption (4.1) we have

$$\begin{aligned} &V((v_i + B_{r_i}) \cap \mathcal{A}) \\ &\quad\subset V(v_i) + r_i\big[\mathcal{E}\big(\mathcal{P}^N(v_i), \alpha(v_i)\big) \oplus B_{\delta(v_i)}\big(\mathcal{Q}^N(v_i)\big)\big] \\ &\quad\subset V(v_i) + r_i\big[\pi^N(i) \oplus B_{\delta_i}^{\perp}\big], \end{aligned} \tag{4.7}$$

where $\pi^N(i) = \pi(\mathcal{P}^N(v_i), \alpha(v_i))$ is the parallelepiped in $\mathcal{P}^N(v_i)H$ with edges of length $2\alpha_k(i) \equiv 2\alpha_k(v_i)$, $i = 1, \ldots, N$ and $B_{\delta_i}^{\perp}$ is the ball $B_{\delta(v_i)}(\mathcal{Q}^N(v_i))$ in $\mathcal{Q}^N(v_i)H$. Cover $\pi^N(i)$ by cubes K^{ij}, $j = 1, 2, \ldots, n(i)$, with edges of length $2\delta_i\epsilon/\sqrt{N}$, $\delta_i = \delta(v_i)$. The diameter of the set $K^{ij} \oplus B_{\delta_i}^{\perp}$ is equal to $2\delta_i\sqrt{1+\epsilon^2} \equiv \gamma_i$. Now, there are $v_{ij} \in \mathcal{A}$ such that $\big[V(u_i) + r_i\big(K^{ij} \oplus B_{\delta_i}^{\perp}\big)\big] \cap \mathcal{A} \subset v_{ij} + B_{\gamma_i r_i}$. The collection of the balls $v_{ij} + B_{\gamma_i r_i}$, $i = 1, \ldots, n(\mathcal{U}_0)$, $j = 1, \ldots, n(i)$, is a new covering $\mathcal{U}_1$ of $\mathcal{A}$.

Let us estimate the number $n(i)$. Obviously,

$$n(i) \leq \prod_{k=1}^{N}\left(\left[\frac{\alpha_k^{(i)}}{\delta_i \epsilon}\sqrt{N}\right]+1\right) \leq C_N^N \frac{\omega_N(i)}{(\delta_i \epsilon)^N}, \tag{4.8}$$
$$C_N = \sqrt{N}+1.$$

Next, the radii of the balls of the covering $\mathcal{U}_1$ are equal to $\gamma_i r_i = 2\delta_i\sqrt{1+\epsilon^2}\, r_i \leq \gamma\rho_0$, where $\gamma = 2\bar{\delta}\sqrt{1+\epsilon^2}$, and $\gamma < 1$ by the choice of ϵ. Hence, we have $\rho_1 \equiv r(\mathcal{U}_1) \leq \gamma\rho_0$. Suppose $\beta \geq N$. Then

$$\begin{aligned} m_{\beta,\,\rho_1}(\mathcal{U}_1) &= \sum_{i=1}^{n(\mathcal{U}_0)}\sum_{j=1}^{n(i)}(\gamma_i r_i)^\beta = \sum_{i=1}^{n(\mathcal{U}_0)}(\gamma_i r_i)^\beta n(i) \\ &\leq C_N^N \overline{\omega}_N \sum_{i=1}^{n(\mathcal{U}_0)} r_i^\beta (2\delta_i\sqrt{1+\epsilon^2})^\beta (\delta_i\epsilon)^{-N} \\ &\leq C_N^N \overline{\omega}_N \left(2\bar{\delta}\sqrt{1+\epsilon^2}\right)^\beta (\bar{\delta}\epsilon)^{-N} \sum_{i=1}^{n(\mathcal{U}_0)} r_i^\beta. \end{aligned}$$

Hence, for $\beta \leq N$

$$m_{\beta,\,\rho_1}(\mathcal{U}_1) \leq C_N^N \overline{\omega} N \left(2\bar{\delta}\sqrt{1+\epsilon^2}\right)^\beta (\epsilon\bar{\delta})^{-N} m_{\beta,\,\rho_0}(\mathcal{U}_0). \tag{4.9}$$

Now, choose β such that

$$\eta = C_N^N \overline{\omega}_N \left(2\bar{\delta}\sqrt{1+\epsilon^2}\right)^\beta (\epsilon\bar{\delta})^{-N} < 1$$

or, equivalently,

$$\beta > N \ln\left(C_N \overline{\omega}_N^{1/N} (\epsilon\bar{\delta})^{-1}\right) \Big/ \ln\left(\frac{1}{2\bar{\delta}\sqrt{1+\epsilon^2}}\right). \tag{4.10}$$

Then, $m_{\beta,\,\rho_1}(\mathcal{U}_1) \leq \eta m_{\beta,\,\rho_0}(\mathcal{U}_0)$ with $\eta < 1$, and $\rho_1 \leq \gamma\rho_0$ with $\gamma < 1$.

Now we start with the covering $\mathcal{U}_1$ and repeat the same procedure to obtain the covering $\mathcal{U}_2$. From $\mathcal{U}_2$ we pass to $\mathcal{U}_3$ and so on. At each step we get $m_{\beta,\,\rho_{k+1}}(\mathcal{U}_{k+1}) \leq \eta m_{\beta,\,\rho_k}(\mathcal{U}_k)$ and $\rho_{k+1} \leq \gamma\rho_k$ with the same η and γ as above. Hence,

$$m_{\beta,\,\rho_k}(\mathcal{U}_k) \leq \eta^k m_{\beta,\,\rho_0}(\mathcal{U}_0),\ \rho_k \leq \gamma^k\rho_0,\ \eta < 1,\ \gamma < 1,$$

for all $k = 1, 2, \ldots$. Therefore, by Lemma 4.1, $\dim_H(\mathcal{A}) \leq \beta$. □

Proof of Theorem 4.2 To prove that $\dim_f(\mathcal{A}) \leq \beta$ it is sufficient to present a sequence of coverings $\mathcal{U}_s$ of the set $\mathcal{A}$ by balls of radius $r_s = r(\mathcal{U}_s)$ such that $r_s \to 0$ and $\nu_\beta(\mathcal{U}_s) := n(\mathcal{U}_s) r_s^\beta \to 0$ when $s \to +\infty$.

Let $\mathcal{U}_0$ be a finite covering of $\mathcal{A}$ by the balls $(v_i + B_{r_0})$, $i = 1, \dots, n(\mathcal{U}_0)$, with $v_i \in \mathcal{A}$. To obtain the covering $\mathcal{U}_1$ we follow the procedure described in the proof of Theorem 4.1, but with certain modifications: the parallelepipeds $\pi^N(i)$ should be covered by cubes K^{ij}, $j = 1, \dots, n(i)$, with edges of length $2\overline{\delta}/\sqrt{N}$ (thus the diameter of $K^{ij} \oplus B^{\perp}_{\delta_i}$ is not greater than $\gamma = 2\sqrt{2\overline{\delta}}$) The sets $\left[V(v_i) + r_0\left(K^{ij} \oplus B^{i\perp}_{\delta_i}\right)\right] \cap \mathcal{A}$ are embedded into the balls $(v_{ij} + B_{\gamma r_0})$. These balls form the covering $\mathcal{U}_1$.

The number $n(i)$, i.e. the number of cubes K^{ij} needed to cover $\pi^N(i)$, is estimated as follows:

$$\begin{aligned} n(i) &\le \prod_{k=1}^{N}\left(\left[\frac{\alpha_k(i)}{\overline{\delta}}\sqrt{N}\right]+1\right) \\ &\le C_N^N \omega_m(i)/\overline{\delta}^{m(i)} \\ &\le C_N^N \overline{\omega}_{m(i)}/\overline{\delta}^{m(i)}, \end{aligned} \tag{4.11}$$

where

$$m(i) = \begin{cases} \max\{k\colon 1 \le k \le N \text{ and } \alpha_k(i) > \overline{\delta}\}, & \text{if } \overline{\delta} < \alpha_1(i), \\ 0 \qquad \text{(in this case } \overline{\omega}_0 := 1), & \text{if } \overline{\delta} \ge \alpha_1(i). \end{cases}$$

It is easy to see that for arbitrary β

$$\begin{aligned} n(\mathcal{U}_1)(r(\mathcal{U}_1))^\beta &\le (r_0\gamma)^\beta \sum_{i=1}^{n(\mathcal{U}_0)} n(i) \\ &\le r_0^\beta C_N^N \sum_{i=1}^{n(\mathcal{U}_0)} \left(2\sqrt{2\overline{\delta}}\right)^\beta \overline{\omega}_{m(i)}/\overline{\delta}^{m(i)} \\ &\le C_N^N \left(2\sqrt{2}\right)^\beta \overline{\delta}^{\beta-N} \chi_N^N n(\mathcal{U}_0) r_0^\beta, \end{aligned} \tag{4.12}$$

where χ_N is defined by (4.5). Now, choose β such that

$$\eta = C_N^N \left(2\sqrt{2}\right)^\beta \overline{\delta}^{\beta-N} \chi_N^N < 1,$$

or equivalently,

$$\beta > N \ln\left(C_N \chi_N \overline{\delta}^{-1}\right) \Big/ \ln\left(\frac{1}{2\sqrt{2\overline{\delta}}}\right). \tag{4.13}$$

Then we get from (4.12): $\nu_\beta(\mathcal{U}_1) \le \eta \nu_\beta(\mathcal{U}_0)$. From the covering $\mathcal{U}_1$ we pass to $\mathcal{U}_2$ and then to $\mathcal{U}_3$ and so on. Obviously, we shall have

$$\nu_\beta(\mathcal{U}_k) \le \eta^k \nu_\beta(\mathcal{U}_0),\ r(\mathcal{U}_k) = \gamma^k r(\mathcal{U}_0) = \gamma^k r_0.$$

Since $\gamma < 1$ and $\eta < 1$, we get

$$\nu_\beta(\mathcal{U}_k) \equiv n(\mathcal{U}_k)(r(\mathcal{U}_k))^\beta \to 0,\ r(\mathcal{U}_k) = \gamma^k r_0 \to 0$$

when $k \to +\infty$, and therefore (4.4) is proved. □

Now we consider the question of verifing the conditions of Theorems 4.1 or 4.2 for some nonlinear differentiable operators $V : H \to H$.

We begin by reminding the reader of some known results concerning linear bounded operators in a Hilbert space.

Let U be a linear bounded operator in H. We associate to U the real numbers

$$\alpha_k(U) := \sup_{\substack{\mathcal{L}\subset H\\ \dim\mathcal{L}=k}} \inf_{\substack{\|x\|=1\\ x\in\mathcal{L}}} \|Ux\|, \tag{4.14}$$

where sup is taken over all linear K-dimensional subspaces $\mathcal{L} \subset H$. The numbers $\alpha_k(U)$ are non-negative defined for all $k = 1, 2, \ldots$, and $\alpha_k(U) \geq \alpha_{k+1}(U)$. Denote $\alpha_\infty(U) := \lim_{k\to\infty} \alpha_k(U)$ and let $\mathcal{T}$ be the set of all indices i such that $\alpha_i(U) > \alpha_\infty(U)$. The set $\mathcal{T}$ may be empty, finite or infinite. Let B_1 be the unit ball in H centered about zero. Then

$$U(B_1) \subset \mathcal{E}(\mathcal{P}^\mathcal{T}(U);\ \alpha(U)) \oplus B_{\alpha_\infty(U)}(\mathcal{Q}^\mathcal{T}(U)), \tag{4.15}$$

where $\mathcal{P}^\mathcal{T}(U)$ is the orthogonal projector on a certain subspace $\mathcal{P}^\mathcal{T}(U)H \subset H$ whose dimension is equal to the number of elements in $\mathcal{T}$, $\mathcal{Q}^\mathcal{T}(U) = I \ominus \mathcal{P}^\mathcal{T}(U)$, $\mathcal{E}(\mathcal{P}^\mathcal{T}(U);\ \alpha(U))$ is an ellipsoid in $\mathcal{P}^\mathcal{T}(U)H$ with the semi-axes $\alpha_i(U)$, $i \in \mathcal{T}$, and $B_{\alpha_\infty(U)}(\mathcal{Q}^\mathcal{T}(U))$ is the ball of radius $\alpha_\infty(U)$ in $\mathcal{Q}^\mathcal{T}(U)H$. From (4.15) we deduce that for any $N < \infty$

$$U(B_1) \subset \mathcal{E}(\mathcal{P}^N(U);\ \alpha(U)) \oplus B_{\sqrt{\alpha_{N+1}^2(U)+\alpha_\infty^2(U)}}(\mathcal{Q}^N(U)) \tag{4.16}$$

where $\mathcal{P}^N(U)$ is the orthogonal projector on a subspace $\mathcal{P}^N(U)H$ of H, $\mathcal{E}(\mathcal{P}^N(U); (U))$ is an ellipsoid of $\mathcal{P}^N(U)H$ with semi-axes $\alpha_1(U), \ldots, \alpha_N(U)$, and $\mathcal{Q}^N(U) = I \ominus \mathcal{P}^N(U)$. Instead of (4.16) we shall use

$$U(B_1) \subset \mathcal{E}(\mathcal{P}^N(U);\ \sqrt{2}\alpha(U)) \oplus B_{\sqrt{2}\alpha_N(U)}(\mathcal{Q}^N(U)). \tag{4.17}$$

We need the following lemma (see [3]):

Lemma 4.2 *Let $\alpha_1, \ldots, \alpha N$, M, k_0 be real numbers satisfying the following conditions: $0 \leq \alpha_N \leq \cdots \leq \alpha_1 \leq M$ and $\omega_N(\alpha) = \alpha_1 \cdots \alpha_N \leq k_0$, where $0 < k_0 \leq M^N$. Let $m := k_0 M^{1-N}$. Define $\alpha'_1, \ldots, \alpha'_N$ according to the rule:*

$$\alpha_i' := \alpha_i \quad \text{if} \quad \alpha_i \geq m, \qquad \text{and} \qquad \alpha_i' := m \quad \text{if} \quad \alpha_i < m.$$

Then $\alpha_i' \geq \alpha_i$, $i = 1, \ldots, N$; $\alpha_N' \geq m$, $\alpha_1' \leq M$, *and* $\omega_N(\alpha') \leq k_0$. *If, in addition,* $\omega_0(\alpha) := 1$ *and for all* $\ell = 0, 1, \ldots, N$ *we have*

$$j_\ell(\alpha) := \omega_\ell(\alpha) k_0^{1-\ell/N} \leq \tilde{k}$$

with some $\tilde{k}$, *then* $j_\ell(\alpha') \leq \tilde{k}$ *for* $\ell = 0, 1, \ldots, N$.

Proof If $\alpha_1 \geq m$ then $\alpha_1' = \alpha_1$, hence $\alpha_1' \leq M$. If $\alpha_1 < m$ then $\alpha_1' < m$. But $m \leq M$ and hence again $\alpha_1' \leq M$. In any case, $\omega_N(\alpha') = \omega_q(\alpha) m^{N-q}$ for some integer q, $0 \leq q \leq N$ ($\omega_0(\alpha) = 1$ by definition). Indeed, if $\alpha_1 \geq \cdots \geq \alpha_q \geq m > \alpha_{q+1} \geq \cdots \geq \alpha_N$ then $\alpha_1' = \alpha_1, \ldots, \alpha_q' = \alpha_q$ and $\alpha_{q+1}' = \cdots = \alpha_N' = m$; if $\alpha_1 \geq \cdots \geq \alpha_N \geq m$ then all $\alpha_i' = \alpha_i$, $j_\ell(\alpha') = j_\ell(\alpha) \leq \tilde{k}$ and $q = N$; if $m > \alpha_1$ then all $\alpha_i' = m$ and $q = 0$.

For $q \leq N - 1$ we have $\omega_N(\alpha') \leq M^q m^{N-q} \leq k_0$. Now consider $j_\ell(\alpha') = \omega_\ell(\alpha') k_0^{1-\ell/N}$ for $q \leq N - 1$. If $\alpha_\ell \geq m$ then $\alpha_1' = \alpha_1, \ldots, \alpha_\ell' = \alpha_\ell$ and $\omega_\ell(\alpha') = \omega_\ell(\alpha)$. Hence $j_\ell(\alpha') = j_\ell(\alpha) \leq \tilde{k}$. If $\alpha_\ell < m$ then $\alpha_1 \geq \cdots \geq \alpha_q \geq m > \alpha_{q+1} \geq \cdots \geq \alpha_\ell$ or $m > \alpha_1$ (and $q = 0$).

In these cases $\omega_\ell(\alpha') = \omega_q(\alpha) m^{\ell-q}$ and therefore

$$\begin{aligned} j_\ell(\alpha') &= \omega_q(\alpha) m^{\ell-q} k_0^{1-\ell/N} = j_q(\alpha)(k_0/M^N)^{(\ell-q)(1-1/N)} \\ &\leq j_q(\alpha) \leq \tilde{k}. \end{aligned}$$

□

Now we return to the embedding (4.17). Because of Lemma 4.2 we have

$$U(B_1) \subset \mathcal{E}(\mathcal{P}^N(U); \sqrt{2}\alpha'(U)) \oplus B_{\sqrt{2}\alpha_N'(U)}(\mathcal{Q}^N(U)), \tag{4.18}$$

with the numbers $\alpha_1'(U), \ldots, \alpha_N'(U)$ computed from $\alpha_1(U), \ldots, \alpha_N(U)$, M and k_0 as was described in Lemma 4.2. Here the only restrictions on M and k_0 are: $M \geq \alpha_1(U)$, $\omega_N(\alpha(U)) \leq k_0$ and $0 < k_0 \leq M^N$.

Let $\mathcal{A}$ be a compact set in H and $V\colon H \to H$ a continuous operator. Suppose that V is *uniformly differentiable on* $\mathcal{A}$, i.e. for every $v \in \mathcal{A}$ there exists a linear bounded operator $U(v)$ such that for all $v_1 \in B_{r_1}(v)$,

$$\|V(v_1) - V(v) - U(v)(v_1 - v)\| \leq \gamma(\|v_1 - v\|)\|v_1 - v\|, \tag{4.19}$$

where $\sup_{v \in \mathcal{A}} \|U(v)\| \leq M < +\infty$ and the function $\gamma(\tau)$ is continuous on the interval $[0, r_1]$ and, $\gamma(0) = 0$. The number r_1 and $\gamma(\tau)$ do not depend on $v \in \mathcal{A}$.

For this linear bounded operator $U(v)$ the numbers $\alpha_k(U(v))$ are computed according to (4.14). We choose M and k_0 so that

$$M \geq \sup_{v \in \mathcal{A}} \alpha_1(U(v)), \quad k_0 \geq \sup_{v \in \mathcal{A}} \omega_N(\alpha(U(v)))$$

and $0 < k_0 \leq M^N$.

Since $\inf_{v \in \mathcal{A}} \alpha'_N(U(v)) \geq m = k_0 M^{1-N}$ there is $r_0 > 0$ such that the r_0-neighbourhood of the set

$$\mathcal{E}(\mathcal{P}^N(U(v)); \sqrt{2}\alpha'(U(v)) \oplus B_{\sqrt{2}\alpha'_N(U(v))}(\mathcal{Q}^N(U(v)))$$

lies entirely in the similar set but with the parameters $2\alpha'_k(U(v))$ instead of $\sqrt{2}\alpha'_k(U(v))$. Thus combining (4.18) and (4.19) we obtain

Lemma 4.3 *Suppose that V is uniformly differentiable on $\mathcal{A}$. Then there is $r_0 > 0$ such that for all $r \leq r_0$ and all $v \in \mathcal{A}$*

$$\begin{aligned} &V((v + B_r) \cap \mathcal{A}) \\ &\quad \subset V(v) + r\big[\mathcal{E}\big(\mathcal{P}^N(U(v)); 2\alpha'(U(v))\big) \oplus B_{2\alpha'_N(U(v))}\big(\mathcal{Q}^N(U(v))\big)\big]. \end{aligned} \tag{4.20}$$

Theorem 4.1 and Lemmas 4.2 and 4.3 yield

Theorem 4.3 *Let V be uniformly differentiable on $\mathcal{A}$, $\mathcal{A} \subset V(\mathcal{A})$ and for differentials $U(v)$, $v \in \mathcal{A}$, of V the inequality*

$$4\sqrt{2}(\sqrt{N} + 1)\overline{k}_0^{1/N} \leq 1, \quad N \geq 1,$$

where $\overline{k}_0 := \sup_{v \in \mathcal{A}} \omega_N(\alpha(U(v))$, holds. Then $\dim_H(\mathcal{A}) \leq N$.

Proof Let k_0 be a positive number in the interval

$$\big[\overline{k}_0, \{4\sqrt{2}(\sqrt{N} + 1)\}^{-N}\big],$$

$M := \max\{k_0^{1/N}; \sup_{v \in \mathcal{A}} \alpha_1(U(v))\}$ and $\alpha'_i(U(v))$, $i = 1, \ldots, N$, be numbers constructed with the help of $\alpha_i(U(v))$, k_0 and M as described in Lemma 4.2. Then the embeddings (4.20) are true and we may use the second statement of Theorem 4.1. In fact, (4.1) holds with $\alpha_i(v) := 2\alpha'_i(U(v))$, $i = 1, \ldots, N$; $\delta(v) := 2\alpha'_N(U(v))$ and $\alpha_1(v) \geq \alpha_2(v) \geq \cdots \geq \alpha_N(v) = \delta(v) > 0$. Moreover, (4.3) is also fulfilled since $\overline{\omega}_N^{1/N} := \sup_{v \in \mathcal{A}} \omega_N^{1/N}(\alpha(v)) \leq 2 \sup_{v \in \mathcal{A}} \omega_N^{1/N}$ $(\alpha'(U(v)) \leq 2k_0^{1/N} \leq \big[2\sqrt{2}(\sqrt{N} + 1)\big]^{-1}$. So $\dim_H(\mathcal{A}) \leq N$. □

Theorem 4.4 *Let V have the same properties as in Theorem 4.3 and suppose that*

$$4\sqrt{2}(\sqrt{N} + 1) \max_{\ell = 0, \ldots, N} \left\{ \sup_{v \in \mathcal{A}} \omega_\ell(\alpha(U(v)))\overline{k}_0^{-1-\ell/N} \right\}^{1/N} \leq 1,$$

where $N \geq 1$, $\overline{k}_0 := \sup_{v \in \mathcal{A}} \omega_N(\alpha(U(v)))$, hold. Then $\dim_f(\mathcal{A}) \leq N$.

Proof If $\overline{k}_0$ is positive we choose $k_0 = \overline{k}_0$,

$$M = \max\{k_0^{1/N};\ \sup_{v\in\mathcal{A}} \alpha_1(U(v))\}$$

and take $\alpha_i'(U(v))$, $i = 1, \dots, N$ corresponding to these numbers and to the numbers $\alpha_i(U(v))$, $i = 1, \dots, N$. Then the embeddings (4.20) are true and we may use the second statement of Theorem 4.2. Actually, (4.1) holds with $\alpha_i(v) := 2\alpha_i'((U(v)), i = 1, \dots, N;\ \delta(v) := 2\alpha_N'(U(v))$ and $\alpha_1(v) \geq \cdots \geq \alpha_N(v) = \delta(v) > 0$. Now we have to verify the condition (4.6) with $\overline{\omega}_\ell = \sup_{v\in\mathcal{A}} \omega_\ell(\alpha(v)) = 2^\ell \sup_{v\in\mathcal{A}} \omega_\ell(\alpha'(U(v)))$ and $\overline{\delta} = \sup_{v\in\mathcal{A}} \delta(v) = 2\sup_{v\in\mathcal{A}} a_N'(U(v))$. According to Lemma 4.2

$$\overline{\delta} \leq 2\sup_{v\in\mathcal{A}} \omega_N^{1/N}(\alpha'(U(v))) \leq 2k_0^{1/N} \text{ and}$$

$$\begin{aligned}\chi_N &:= \max_{\ell=0,\,1,\dots,\,N} \overline{\omega}_\ell^{1/N}\overline{\delta}^{1-\ell/N}\\ &\leq 2\sup_{v\in\mathcal{A}}\{\max_{\ell=0,\dots,\,N} \omega_\ell(\alpha'(U(v)))k_0^{1-\ell/N}\}^{1/N}\\ &\leq 2\sup_{v\in\mathcal{A}}\{\max_{\ell=0,\dots,\,N} \omega_\ell(\alpha(U(v)))k_0^{1-\ell/N}\}^{1/N}.\end{aligned}$$

The inequality $2\sqrt{2}\big(\sqrt{N}+1\big)\chi_N \leq 1$ is therefore guaranteed by the condition of Theorem 4.4 and so $\dim_f(\mathcal{A}) \leq N$.

If $\overline{k}_0 = 0$ we take $M = \max\{1; \hat{\alpha}_1\}$, where $\hat{\alpha}_1 := \sup_{v\in\mathcal{A}} \alpha_1(U(v))$, and a number $k_0 \in \big(0, (\beta M)^{-N^2}\big]$, where $\beta := 4\sqrt{2}\big(\sqrt{N}+1\big)$, and construct $\alpha_i'(U(v)), i = 1, \dots, N$, the corresponding $\alpha_i'(U(v)), i = 1, \dots, N$, and these k_0 and M.

Then

$$\begin{aligned}\beta^{-N} &\geq \max_{\ell=0,\dots,\,N-1}\{M^\ell k_0^{1-\ell/N}\}\\ &\geq \max_{\ell=0,\dots,\,N}\{\sup_{v\in\mathcal{A}} \omega_\ell(\alpha(U(v)))k_0^{1-\ell/N}\}.\end{aligned}$$

According to Lemma 4.2

$$\max_{\ell=0,\dots,\,N}\{\omega_\ell(\alpha(U(v)))k_0^{1-\ell/N}\} \geq \max_{\ell=0,\dots,\,N}\{\omega_\ell(\alpha'(U(v)))k_0^{1-\ell/N}\},$$

for all $v \in \mathcal{A}$.

Therefore

$$\begin{aligned}\chi_N &= \max_{\ell=0,\dots,N}\{\overline{\omega}_N^{1/N}\overline{\delta}^{1-\ell/N}\}\\ &\le 2\max_{\ell=0,\dots,N}\{\sup_{v\in\mathcal{A}}\omega_\ell(\alpha'(U(v)))k_0^{1-\ell/N}\}^{1/N}\\ &\le\left[2\sqrt{2}(\sqrt{N}+1)\right]^{-1}.\end{aligned}$$

So the condition (4.6) is fulfilled and $\dim_f(\mathcal{A})\le N$. □

When $\mathcal{A}$ is invariant, i.e. $V(\mathcal{A})=\mathcal{A}$, the following results hold.

Theorem 4.5 *Suppose that V is a uniformly differentiable operator on a compact invariant set $\mathcal{A}$, and its differentials $U(v)$, $v\in\mathcal{A}$, satisfy the condition:*

$$\sup_{v\in\mathcal{A}}\omega_N(\alpha(U(v))) := \overline{k}_0 < 1. \tag{4.21}$$

Then $\dim_H(\mathcal{A})\le N$.

Theorem 4.6 *Under the assumptions of Theorem 4.5 let*

$$\max_{\ell=0,\dots,N}\{\sup_{v\in\mathcal{A}}\omega_\ell(\alpha(U(v))))\overline{k}_0^{1-\ell/N}\} < 1, \tag{4.22}$$

where $\omega_0(\alpha):=1$.

Then $\dim_f(\mathcal{A})\le N$.

Proof of Theorem 4.5 Theorem 4.5 is deduced from Theorem 4.3 applied to the operator V_p the p-th power of the operator V which is denoted now by V_1.

It is known that $\omega_k(\alpha(U_1\cdot U_2))\le\omega_k(\alpha(U_1))\omega_k(\alpha(U_2))$ for all $k\ge 1$ and arbitrary bounded linear operators U_1, U_2. Since the differentials $U_p(v)$, $v\in\mathcal{A}$ of the operators V_p have the form $U(v_1)\dots U(v_p)$, where $v_k\in\mathcal{A}$, and $U(v_k)$ is the differential of V_1, the following inequalities hold:

$$\sup_{v\in\mathcal{A}}\omega_k(\alpha(U_p(v))) := \overline{\omega}_k(p)\le(\overline{\omega}_k(1))^p.$$

By the assumption (4.21) we have $\overline{\omega}_N(p)\le\overline{k}_0^p$. Choose the integer p large enough that $4\sqrt{2}(\sqrt{N}+1)\overline{k}_0^{p/N}\le 1$. Applying Theorem 4.3 (for the operator $V=V_p$) we obtain $\dim_H(\mathcal{A})\le N$. □

In a similar way Theorem 4.6 is deduced from Theorem 4.4.

Theorems 4.5 and 4.6 may be used to estimate the Hausdorff and fractal dimensions of compact sets $\mathcal{A}$ which are invariant with respect to evolution operators $V_t : H \to H, t \in \mathbb{R}^+$, of the problems

$$\partial_t v(t) = \Phi(v(t)), \quad v|_{t=0} = v_0, \tag{4.23}$$

where Φ is a nonlinear (generally unbounded) operator, subjected to some restrictions.

Suppose we know that the problem (4.23), for every $v_0 \in H$, has a unique solution defined on $\mathbb{R}^+$, and the solution operators V_t, $t \in \mathbb{R}^+$, form a semigroup in H. Let $\mathcal{A}$ be a compact set invariant with respect to this semigroup. In what follows we deal exclusively with the semigroup V_t restricted to the set $\mathcal{A}$.

Let $U(t, v_0)$ be the differential of the operator V_t at a point $v_0 \in H$. It is known that $U(t, v_0)$ is the solution operator for the linear problem

$$\partial_t u(t) = L(t)u(t), \quad u|_{t=0} = u_0, \tag{4.24}$$

where $L(t) = L(t, v_0)$ is the differential of $\Phi(.)$ at the point $v(t) = V_t(v_0)$.

Usually the investigation of the nonlinear problem (4.23) begins with the study of some linearizations of problem (4.23) similar to (4.24). As a final result one gets the existence of the operators $V_t, t \in \mathbb{R}^+$, and of its differentials. In Part II we shall consider some problems of this kind.

Now we are interested in estimating from above the numbers ω_n for the operators $U(t, v_0)$ We consider two classes of problem (4.23). For the first class, the operators V_t and $U(t, v_0)$ will be compact $t > 0$; this very property distinguishes the problems of parabolic type. For the second class, these operators are merely continuous and bounded (for example, in the case of problems of hyperbolic type). In either case, the numbers ω_n may be estimated as follows.

It is known that for every linear bounded operator U, the number $\omega_n(U) \equiv \omega_n(\alpha(U)) = \alpha_1(U) \cdot \cdots \cdot \alpha_n(U)$ (where $\alpha_k(U)$ are defined in (4.14)) is the norm of the operator $\Lambda^n(U)$ in the Hilbert space $\Lambda^n(H)$. The space $\Lambda^n(H)$ consists of the elements $u_1 \wedge \cdots \wedge u_n$ with $u_k \in H$, and the inner product in $\Lambda^n(H)$ is defined by $(u_1 \wedge \cdots \wedge u_n, v_1 \wedge \cdots \wedge v_n) = \det((u_i, v_j))$, where $(\cdot, \cdot)$ is the inner product in H. We denote the norm of $u_1 \wedge \cdots \wedge u_n$ by $\|u_1 \wedge \cdots \wedge u_n\|$.

The operator $\Lambda^n(U)$ acts in $\Lambda^n(H)$ according to the rule

$$\Lambda^n(U)(u_1 \wedge \cdots \wedge u_n) = Uu_1 \wedge \cdots \wedge Uu_n.$$

It is known also (see [9], [10], etc.) that

$$\begin{aligned}\sum_{k=1}^{n}(u_1\wedge\cdots\wedge Uu_k\cdots\wedge u_n,\quad u_1\wedge\cdots\wedge u_k\cdots\wedge u_n)\\ =\|u_n\wedge\cdots\wedge u_n\|^2\mathrm{tr}(\mathcal{P}^n(\vec{u})U\mathcal{P}^n(\vec{u})),\end{aligned}\tag{4.25}$$

where $\mathcal{P}^n(\vec{u})$ is the orthogonal projector on the span $\{u_1,\ldots,u_n\}$ in H. Since $\omega_n(U)$ is the norm of $\Lambda^n(U)$ in $\Lambda^n(H)$, we have $\omega_n(U_1U_2)\le\omega_n(U_1)\omega_n(U_2)$; this inequality was used in the proofs of Theorems 4.5 and 4.6.

Let us take n linearly independent elements $u_i(0)\in H$ and let $u_i(t)$ be the corresponding solutions of (4.24) with $u_i|_{t=0}=u_i(0)$. Obviously, $u_i(t)$ are linearly independent for all $t>0$.

Let $\mathcal{P}^n(\vec{u}(t)):=\mathcal{P}^n(t)$ be the orthogonal projector on the span of $\{(u_1(t),\ldots,u_n(t)\}$. From (4.24) we get

$$\partial_t u_i(t)=L_{(n)}(t)u_i(t)+\partial_t\mathcal{P}^n(t)u_i(t),\quad i=1,\ldots,n,\tag{4.26}$$

where $L_{(n)}(t)=\mathcal{P}^n(t)L(t)\mathcal{P}^n(t)$. Set $G_n(t):=\|u_1(t)\wedge\cdots\wedge u_n(t)\|^2$. By (4.26) we have the Liouville formula:

$$\begin{aligned}\frac{1}{2}\partial_t G_n(t)&=G_n(t)\,\mathrm{tr}\big[L_{(n)}(t)+\partial_t\mathcal{P}^n(t)\big]\\ &=G_n(t)\,\mathrm{tr}L^c_{(n)}(t),\end{aligned}\tag{4.27}$$

where $L^c_{(n)}(t)=\frac{1}{2}[L_n(t)+L^*_n(t)]$. Here we have used the identities $\mathrm{tr}(\partial_t\mathcal{P}^n(t))=\partial_t\,\mathrm{tr}\,\mathcal{P}^n(t)$ and $\mathrm{tr}\,\mathcal{P}^n(t)=n$, valid for any orthogonal projector $\mathcal{P}^n(t)$.

Formula (4.27) is proved by the same procedure as the usual Liouville formula for a system of n ordinary differential equations in $\mathbb{R}^n$. (Although the system (4.26) is not a system of ODE, since $\partial_t u_i(t)$ and $\partial_t\mathcal{P}^n(t)u_j(t)$ might not lie in $\mathcal{P}^n(t)H$.) Integrating (4.27) we obtain

$$\|u_1(t)\wedge\cdots\wedge u_n(t)\|=\|u_1(0)\wedge\cdots\wedge u_n(0)\|\exp\left\{\int_0^t\mathrm{tr}\big[L^c_{(n)}(\tau)\big]d\tau\right\},\tag{4.28}$$

where $u_i(t)=U(t)u_i(0)$ and $U(t)$ is the solution operator for (4.24). Hence the following estimate holds:

$$\begin{aligned}\omega_n(U(t))&=\|\Lambda^n(U(t))\|\\ &=\sup_{u_k(0)\in H}\frac{\|U(t)u_1(0)\wedge\cdots\wedge U(t)u_n(0)\|}{\|u_1(0)\wedge\cdots\wedge u_n(0)\|}\\ &\le\sup_{u_k(0)\in H}\exp\left\{\int_0^t\mathrm{tr}\big(\mathcal{P}^n(\tau)L^c_{(n)}(\tau)\mathcal{P}^n(\tau)\big)d\tau\right\},\end{aligned}\tag{4.29}$$

where $\mathcal{P}^n(t)=\mathcal{P}^n(\vec{u}(t))$ is the orthogonal projector on the span $\{(U(t)u_1(0),\ldots,U(t)u_n(0)\}$.

Usually when the existence problem for (4.24) is investigated, the "strength" of the operators $L(t)$ is "measured" by some fixed self-adjoint positively defined operator A, whose inverse A^{-1} is compact. Here we shall follow the same line.

Let H_s, $s \in \mathbb{R}$, be a scale of Hilbert spaces with the inner products $(u,v)_s \equiv (A^{s/2}u, A^{s/2}v)$ and the norms $\|u\|_s = (u,u)_s^{1/2}$. The original H coincides with $H_0, (\cdot,\cdot)_0 = (\cdot,\cdot)$ and $\|\cdot\|_0 = \|\cdot\|$.

For some problems of parabolic type the operators $A^{-1/2}L(t)A^{-1/2}$ are bounded and for almost all $t \geq 0$

$$(L^c(t)u,u) \leq -h_1(t)\|u\|_1^2 + \sum_{k=1}^{m} h_{s_k}(t)\|u\|_{s_k}^2, \tag{4.30}$$

for any $u \in H_1$, where $s_k \leq 0$, h_1, $h_{s_k} \in L_{1,loc}(\mathbb{R})$ and $h_{s_k}(t) \geq 0$, $h_1(t) \geq h_1 > 0$. These inequalities make it possible to estimate the right hand side of (4.29).

Explicitly, if $\mathcal{P}^n$ is an orthogonal projector on the span $\{\phi_1, \ldots, \phi_n\}$, where $\phi_i \in H$ and $(\phi_i, \phi_j) = \delta_{ij}$, the following estimates hold:

$$\begin{aligned}\operatorname{tr}(\mathcal{P}^n L^c(t)\mathcal{P}^n) &= \sum_{i=1}^{n}(L^c(t)\phi_i,\phi_i)\\ &\leq -h_1(t)\sum_{i=1}^{n}(A\phi_i,\phi_i) + \sum_{k=1}^{m} h_{s_r}(t)\sum_{i=1}^{n}(A^{s_k}\phi_i,\phi_i).\end{aligned} \tag{4.31}$$

Since A is self-adjoint with positive eigenvalues $\lambda_k(A)$ and $\lambda_k(A) \nearrow +\infty$ as $k \to +\infty$, in view of the min-max principle we have

$$\sum_{i=1}^{n}(A^s\phi_i,\phi_i) \leq \mathrm{Sp}_n A^s, \quad \text{if } s \leq 0, \quad \text{and}$$

$$\sum_{i=1}^{n}(A^s\phi_i,\phi_i) \geq \mathrm{Sp}_n A^s, \quad \text{if } s \geq 0.$$

Therefore, (4.31) implies

$$\operatorname{tr}(\mathcal{P}^n L^c(t)\mathcal{P}^n) \leq -h_1(t)\,\mathrm{Sp}_n A + \sum_{k=1}^{m} h_{s_k}(t)\,\mathrm{Sp}_n A^{s_k}, s_k \leq 0. \tag{4.32}$$

For the class of problems under consideration (problems of parabolic type), the solutions $u_k(t) = U(t)u_k(0)$, $u_k(0) \in H$, for almost all $t \in \mathbb{R}^+$ lie in H_1, $\int_0^t \|u_k(\tau)\|_1^2 d\tau < +\infty$ and $u_k \in C(\mathbb{R}^+, H)$. Hence $\mathcal{P}^n(\vec{u}(t))$ satisfies for almost all t the assumptions used in the proof of (4.32). Therefore, (4.32) and (4.29) yield

$$\omega_n(U(t)) \le \exp\left\{-\int_0^t h_1(\tau)\mathrm{d}\tau \mathrm{Sp}_n A + \sum_{k=1}^m \int_0^t h_{s_k}(\tau)\mathrm{d}\tau \mathrm{Sp}_n A^{s_k}\right\} \tag{4.33}$$

for all $t \in \mathbb{R}^+$. The terms with $s_k \neq 0$ appear in some applications.

As for problems of hyperbolic type (and many others) the operators V_t and $U(t, v_0)$ are merely continuous and bounded. Instead of (4.30) in that case, for any $u \in H$, we have:

$$(L^c(t)u, u) \le -h_0(t)\|u\|^2 + \sum_{k=1}^m h_{s_k}(t)\|u\|_{s_k}^2, \tag{4.34}$$

where $s_k < 0$; $h_0, h_{s_k} \in L_{1,loc}(\mathbb{R})$, $h_{s_k}(t) \ge 0$, $h_0(t) \ge h_0 > 0$.

As in the proof of (4.33) we obtain

$$\omega_n(U(t)) \le \exp\left\{-n\int_0^t h_0(\tau)\mathrm{d}\tau + \sum_{k=1}^m \int_0^t h_{s_k}(\tau)\mathrm{d}\tau \mathrm{Sp}_n A^{s_k}\right\}. \tag{4.35}$$

We now summarize the results:

Theorem 4.7 *Let the operators $L(t)$ in* (4.24) *satisfy the inequalities* (4.30) *or* (4.34)*. Then the corresponding estimates* (4.33) *or* (4.35) *(for $t \ge 0$ and $n \ge 1$) hold.*

From Theorems 4.5 and 4.7 follows:

Theorem 4.8 *Let $\{V_t,\ t \in \mathbb{R}^+,\ H\}$, be a semigroup of solution operators for problem* (4.23)*, and $\mathcal{A}$ a compact set invariant with respect to V_t. Let V_t and $\Phi(\cdot)$ be uniformly differentiable on $\mathcal{A}$ and let $L(t, v_0)$ be a differential of Φ at the point $V_t(v_0)$, $v_0 \in \mathcal{A}$. Suppose that $L(t, v_0)$, $v_0 \in \mathcal{A}$, satisfies the inequalities* (4.30) *or* (4.34) *with the functions $h_i(t)$ independent of $v_0 \in \mathcal{A}$.*

Then $\dim_H(\mathcal{A}) \le N$ *where N is the minimal positive integer such that the expression in the braces on the right hand side of* (4.33) *(respectively,* (4.35)*) is negative for some $t > 0$.*

In order to majorize $\dim_f(\mathcal{A})$ we have to estimate from above

$$j_n(t) \equiv \overline{\omega}_n(t)\overline{\omega}_N^{1-n/N}(t), \quad n = 0, 1, \ldots, N,$$

where $\overline{\omega}_n(t) \equiv \sup_{v_0 \in \mathcal{A}} \omega_n(U(t, v_0))$. Denote by $U(t)$ the operator $U(t, v_0)$ for some $v_0 \in \mathcal{A}$, and $\overline{h}_s(t) := t^{-1}\int_0^t h_s(\tau)\mathrm{d}\tau$. Suppose that

$$\begin{aligned} \mathrm{Sp}_n A &\ge c_1 n^{1+\gamma}, \qquad \gamma > 0, \quad c_1 > 0, \\ \mathrm{Sp}_n A^s &\le c_s n^{1+s}, \qquad s \le 0, \quad c_0 = 1. \end{aligned} \tag{4.36}$$

Such inequalities hold for the Laplace operator, for the Stokes operator (with any of the classical boundary conditions) and for many other operators in mathematical physics.

Take, for example, one of the summands from $\sum_{k=1}^{m}$ in (4.30) (or (4.34)), say $h_s(t)\|u\|_s^2$, $s \leq 0$. From (4.36) it follows that, for each $n \geq 1$,

$$\begin{aligned} \omega_n(U(t)) &\leq \exp t\big[-\overline{h}_1(t)c_1 n^{1+\gamma} + \overline{h}_s(t)c_s n^{1+s}\big] \\ &\equiv \Psi_n(t). \end{aligned} \tag{4.37}$$

This $\Psi_n(t)$ majorizes $\overline{\omega}_n(t)$ as well. Therefore,

$$\begin{aligned} j_n(t) &\equiv \overline{\omega}_n(t)\overline{\omega}_N^{1-n/N}(t) \leq \Psi_n(t)\Psi_N(t)^{1-n/N}, \\ \frac{1}{t}\ln j_n(t) &\leq \frac{1}{t}\left[\ln \Psi_n(t) + \left(1-\frac{n}{N}\right)\ln \Psi_N(t)\right] \\ &\equiv -\overline{h}_1(t)c_1 n^{1+\gamma} + \overline{h}_s(t)c_s n^{1+s} \\ &\quad + \left(1-\frac{n}{N}\right)\big[-\overline{h}_1(t)c_1 N^{1+\gamma} + \overline{h}_s(t)c_s N^{1+s}\big] \\ &= -\overline{h}_1(t)c_1\big(n^{1+\gamma} + N^{1+\gamma} - nN^{\gamma}\big) \\ &\quad + \overline{h}_s(t)c_s\big(n^{1+s} + N^{1+s} - nN^{s}\big). \end{aligned}$$

By the Young inequality we have

$$\begin{aligned} n^{1+\gamma} + N^{1+\gamma} - nN^{\gamma} &\geq n^{1+\gamma} + N^{1+\gamma} - \frac{1}{1+\gamma}n^{1+\gamma} - \frac{\gamma}{1+\gamma}N^{1+\gamma} \\ &= \frac{1}{1+\gamma}N^{1+\gamma} + \frac{\gamma}{1+\gamma}n^{1+\gamma} \geq \frac{1}{1+\gamma}N^{1+\gamma}, \end{aligned}$$

for $\gamma > 0$; equality also holds in the case $\gamma = 0$.

The second sum is estimated from above for $n \in [1, N]$ (remember that $s \leq 0$):

$$n^{1+s} + N^{1+s} - nN^{s} \leq N^{1+s}.$$

Thus we obtain

$$\frac{1}{t}\ln j_n(t) \leq -\overline{h}_1(t)c_1\frac{1}{1+\gamma}N^{1+\gamma} + \overline{h}_s(t)c_s N^{1+s}, \quad 1 \leq n \leq N-1, \tag{4.38}$$

and besides,

$$\begin{aligned} \frac{1}{t}\ln j_N(t) = \frac{1}{t}\ln j_0(t) &= \frac{1}{t}\ln \Psi_N(t) \\ &\equiv -\overline{h}_1(t)c_1 N^{1+\gamma} + \overline{h}_s(t)c_s N^{1+s}. \end{aligned} \tag{4.39}$$

In view of Theorem 4.8, $\dim_H(\mathcal{A}) \leq N$, where N has been chosen in such a way that

$$-\overline{h}_1(t)c_1N^{\gamma-s} + \overline{h}_s(t)c_s < 0 \tag{4.40}$$

for some $t > 0$.

In order to get $\dim_f(\mathcal{A}) \leq N$ (see Theorem 4.6) it is sufficient to choose t and N so that

$$-\overline{h}_1(t)c_1\frac{1}{1+\gamma}N^{\gamma-s} + \overline{h}_s(t)c_s < 0. \tag{4.41}$$

Remember that $\gamma > 0$ and $s \leq 0$ in (4.40) and (4.41) (see (4.36)).

The above calculations also show that in the case (4.34) $\dim_H(\mathcal{A}) \leq \dim_f(\mathcal{A}) \leq N$, where N is chosen in such a way that

$$-\overline{h}_0(t)N^{-1} + \overline{h}_s(t)c_s < 0, \quad s < 0$$

for some $t > 0$.

When there are several summands in $\sum_{k=1}^{m}$, we proceed in the same fashion and obtain:

Theorem 4.9 *Under the assumptions of Theorem 4.8, suppose that the inequalities* (4.36) *hold. If $L(t)$ satisfies the condition* (4.30), *then $\dim_H(\mathcal{A}) \leq N$, where N is such that*

$$-\overline{h}(t)c_1N^{\gamma} + \sum_{k=1}^{m}\overline{h}_{s_k}(t)c_{s_k}N^{s_k} < 0,\ \gamma > 0,\ s_k \leq 0, \tag{4.42}$$

for some $t > 0$. If N is such that

$$-\frac{1}{1+\gamma}\overline{h}_1(t)c_1N^{\gamma} + \sum_{k=1}^{m}\overline{h}_{s_k}(t)c_{s_k}N^{s_k} < 0, \quad \gamma > 0,\ s_k \leq 0 \tag{4.43}$$

for some $t > 0$, then $\dim_f(\mathcal{A}) \leq N$.

If $L(t)$ satisfies the condition (4.34), *then $\dim_H(\mathcal{A}) \leq \dim_f(\mathcal{A}) \leq N$, where N is such that*

$$-\overline{h}_0(t) + \sum_{k=1}^{m}\overline{h}_{s_k}(t)c_{s_k}N^{s_k} < 0, \quad s_k < 0 \tag{4.44}$$

for some $t > 0$.

Usually in some problems of mathematical physics (e.g. for the Navier–Stokes equations) the functions $\overline{h}_i(t) \equiv \frac{1}{t}\int_0^t h_i(\tau)\mathrm{d}\tau$ decrease when $t \to +\infty$, and $h_1(t)$ in (4.30) or $h_0(t)$ in (4.34) does not depend on t. In this case we may put $\overline{h}_{s_k}(\infty)$ instead of $\overline{h}_{s_k}(t)$ in (4.41)–(4.44).

PART II

Semigroups generated by evolution equations

5

Introduction to Part II

In Part II we consider abstract semi-linear evolution equations mainly of hyperbolic type. They generate semigroups of class $A\mathcal{K}$. Evolution equations of parabolic type generate semigroups of class. $\mathcal{K}$. We devote to them only the short Chapter 6, as semi-linear parabolic equations are expounded in a comparatively complete literature. Many publications are devoted to the Navier–Stokes equations which generate (in the two-dimensional case) semigroups of class $\mathcal{K}$. In the first publication on this subject [1] the set $\mathcal{M}$ of all limit states or, which is the same, the minimal global B-attractor was found; among its properties the most interesting one is a finiteness of the dynamics $\{V_t\}$ on $\mathcal{M}$ (or alternatively, the finiteness of the number N_1 of determining modes). Here we give some comparatively recent results [11] concerning majorants for the number N_1 and for the fractal dimension of invariant bounded sets which are better (for small viscosity ν) than before. On the other hand, quasi-linear parabolic equations of general form also generate semigroups of class $\mathcal{K}$, but the presentation of this material requires a separate publication. For this purpose we need to use results on the global unique solvability of boundary value problems for these equations and estimates of a local type. They may be found in the monograph [15]; more recent results are described in the survey [16], which also contains a list of publications. It is, nevertheless, necessary to put this subject in the framework of the theory of semigroups choosing the phase spaces correctly.

We describe now some known general facts used in the next chapters.

Let H be a separable Hilbert space, $(\cdot,\cdot)$ and $\|\cdot\|$ denote the inner product and norm in H, and A a linear unbounded operator with domain $\mathcal{D}(A)$ dense in H. Moreover, A is self-adjoint, positive definite and its inverse A^{-1} is completely continuous. Let us denote by $0 < \lambda_1 \leq \lambda_2 \leq \cdots$ the eigenvalues of A and by $\phi_1, \phi_2, \ldots$ the corresponding orthonormalized eigenelements.

Starting from H and A we construct by the usual procedure the space-scale $H_s(A)$, $s \in \mathbb{R}$: $H_s(A)$ is the domain $\mathcal{D}(A^{s/2})$ of $A^{s/2}$ with the inner product

$$(u, v)_{s, A} = (A^{s/2}u, A^{s/2}v) \tag{5.1}$$

and the norm $\|\cdot\|_{s, A}$. Clearly, $H_0(A) \equiv H$. The spaces $H_s(A)$ and $H_{-s}(A)$ are dual with respect to H. We simply write:

$$(u, v) = (A^{s/2}u, A^{-s/2}v) \tag{5.2$_1$}$$

if $u \in H_s(A)$ and $v \in H_{-s}(A)$, and

$$\begin{aligned}(u, v)_{r, A} &= (A^{s/2}u, A^{-s/2}v)_{r, A}\\ &= (A^{(r+s)/2}u, A^{(r-s)/2}v)\end{aligned} \tag{5.2$_2$}$$

if $u \in H_{s+r}(A)$ and $v \in H_{-s+r}(A)$. It is easy to verify that

$$\begin{gathered}\|u\|^2_{s, A} \geq \lambda_1^{\delta}\|u\|^2_{s-\delta, A} \quad \text{for} \quad \delta > 0,\ u \in H_s(A),\\ A^{\delta/2}(H_{s-\delta}(A)) = H_s(A), \quad \left\|A^{\delta/2}u\right\|_{s.A} = \|u\|_{s+\delta, A},\\ (u, v)_{s+\delta, A} = (A^{\delta}u, v)_{s, A},\\ |(u, v)_{s, A}| \leq \|u\|_{s+\delta, A}\|v\|_{s-\delta, A}, \quad \text{etc.}\end{gathered} \tag{5.3}$$

In Chapter 6 we use only the scale $H_s(A)$, $s \in \mathbb{R}$, with a fixed operator A and omit A in the notations $H_s(A)$, $(\cdot, \cdot)_{s, A}$ and $\|\cdot\|_{s, A}$. But the dependence on A will be explicitly indicated in Chapter 7, where we deal with operators depending on a parameter. We shall denote the set of all linear bounded operators acting from the Banach space X to the Banach space Y by the symbol $L(X \to Y)$, their norms by the symbol $\|\cdot\|_{L(X \to Y)}$.

6

Estimates for the number of determining modes and the fractal dimension of bounded invariant sets for the Navier–Stokes equations

It is well known that the Navier–Stokes equations with some boundary conditions can be considered as the equation

$$\partial_t v(t) + \nu A v(t) + f(v(t)) = h \tag{6.1}$$

in a Hilbert space H. In the case of homogeneous no-slip boundary conditions (i.e. $v(x,t)|_{x\in\partial\Omega} = 0$ where Ω is a bounded domain $\mathbb{R}^m$, $m = 2$ or 3, with a smooth boundary $\partial\Omega$), $H \equiv \overset{\circ}{\mathcal{J}}(\Omega)$ is a subspace of the vector-space $L_2^m(\Omega)$. In the case of periodic boundary conditions (here Ω is a parallelepiped) H is another subspace of $L_2^m(\Omega)$. In both cases A enjoys the properties indicated in Chapter 5 (like the scalar Laplace operator $(-\Delta)$ in $L_2(\Omega)$ with the boundary condition $u|_{\partial\Omega} = 0$). Elements v of H are vector functions $v\colon \boldsymbol{x} \in \Omega \to v(\boldsymbol{x}) = (v_1(\boldsymbol{x}), \ldots v_m(\boldsymbol{x})) \in \mathbb{R}^m$ and $f(v) = P\mathcal{F}(v)$ where P is the orthoprojector of $L_2^m(\Omega)$ onto H and $\mathcal{F}$ is defined by the mapping:

$$\mathcal{F}(v)\colon v \to \sum_{k=1}^{m} v_k(x)\frac{\partial}{\partial_{x_k}} v(\boldsymbol{x}) \in \mathbb{R}^m, \quad x \in \Omega.$$

In [1] it is proved that in the case $m = 2$ (for two-dimensional Ω) the solution operators V_t of problem (6.1) with $h \in H$ form a continuous semigroup $\{V_t, t \in \mathbb{R}^+, H\}$ of class $\mathcal{K}$ and this semigroup has a compact connected minimal global B-attractor $\mathcal{M}$, lying in the ball $B_{R_0} = \{v\colon \|v\| \le R_0 = \|h\|(\nu\lambda_1)^{-1}\}$. The set $\mathcal{M}$ is bounded in the space H_2. It is also maximal among all bounded invariant sets of our semigroup.

The semigroup can be extended on $\mathcal{M}$ to a continuous group $\{V_t, t \in \mathbb{R}^+, \mathcal{M}\}$, and, what is more, this group is in some sense finite-dimensional. It means that there is an integer N_1 such that the projection $P^{N_1}\gamma(v)$ of any complete trajectory $\gamma(v)$, $v \in \mathcal{M}$ in the subspace span $\{\phi_1, \ldots, \phi_{N_1}\} \equiv P^{N_1}H$ determines the trajectory $\gamma(v)$ (here ϕ_k, $k = 1, \ldots, N_1$, are eigenelements of A). More precisely, if for $\gamma(v)=\{V_t(v),\, t\in\mathbb{R}\}$ and $\gamma(\tilde v)=\{V_t(\tilde v),\, t\in\mathbb{R}\}$ with $v,\tilde v\in\mathcal{M}$ we have the equalities $P^{N_1}V_t(v)=P^{N_1}V_t(\tilde v)$ for all $t\in\mathbb{R}$, then $V_t(v)=V_t(\tilde v)$ for all $t\in\mathbb{R}$.

We call the smallest of such integer numbers the *number* N_1 *of determining modes for* $\mathcal{M}$ (or for $\mathcal{A}$ if we consider another invariant set $\mathcal{A}$). In [1] a majorant for N_1 is computed. Here we shall deduce a majorant which is better for small ν^{-1}. Let us remark that we do not use the fact that $\mathcal{M}$ is an attractor.

For an arbitrary complete trajectory $v(t)$, $t\in\mathbb{R}$ lying in $\mathcal{M}$ we have the estimates:

$$\sup_{t\in R}\|v(t)\|\le R_0\equiv\|h\|(\nu\lambda_1)^{-1} \tag{6.2}$$

and

$$\nu\int_\tau^t\|v(\xi)\|_1^2\mathrm{d}\xi\le\frac{1}{2}R_0^2+\|h\|^2(\lambda_1\nu)^{-1}|t-\tau|, \tag{6.3}$$

where λ_k, $k=1,2,\ldots$, are the eigenvalues of A. The difference $u(t)=v(t)-\tilde v(t)$ of two solutions of (6.1) satisfies the equality:

$$\partial_t u(t)+\nu Au(t)+P(\tilde v_k(t)u_{x_k}(t)+u_k(t)v_{x_k}(t))=0,\ t\in\mathbb{R}. \tag{6.4}$$

From (6.4) we deduce

$$\begin{aligned}
&\frac{1}{2}\frac{\mathrm{d}}{\mathrm{d}t}\|u(t)\|^2+\nu\|u(t)\|_1^2\\
&\quad=-\int_\Omega u_k(t,x)v_{x_k}(t,x)u(t,x)\mathrm{d}x\\
&\quad\le\|v(t)\|_1\|u(t)\|_{L_4(\Omega)}^2\le\frac{1}{\sqrt{2}}\|v(t)\|_1\|u(t)\|_1\|u(t)\|\\
&\quad\le\frac{\nu}{2}\|u(t)\|_1^2+\frac{1}{4\nu}\|v(t)\|_1^2\|u(t)\|^2,
\end{aligned}$$

and from this inequality it follows that

$$\frac{d}{dt}\|u(t)\|^2+\nu\|u(t)\|_1^2\le(2\nu)^{-1}\|v(t)\|_1^2\|u(t)\|^2. \tag{6.5}$$

Suppose that $P^Nu(t)=0$ for all $t\in\mathbb{R}$. Then $u(t)=Q^Nu(t)$ where $Q^N:=I-P^N$ and

$$\|u(t)\|_1^2\ge\lambda_{N+1}\|Q^Nu(t)\|^2 \tag{6.6}$$

(as well as for an arbitrary element of Q^NH), so that

$$\frac{\mathrm{d}}{\mathrm{d}t}\|u(t)\|^2\le-h(t)\|u(t)\|^2, \tag{6.7}$$

with $h(t) = \nu\lambda_{N+1} - (2\nu)^{-1}\|v(t)\|_1^2$. If $\int_\tau^t h(\xi)\mathrm{d}\xi \to +\infty$ when $\tau \to -\infty$ then $u(t) = 0$ for all $t \in \mathbb{R}$. By (6.3) this will hold if

$$\lambda_{N+1} > \nu^{-4}\|h\|^2(2\lambda_1)^{-1}. \tag{6.8}$$

As $\lambda_k = O(k)$ we can satisfy (6.8) by choosing an N which satisfies the inequality

$$N \le c_1\nu^{-4} + c_1', \text{ with some } c_1, c_1' \in \mathbb{R}^+. \tag{6.9}$$

In the case of periodic boundary conditions we have the following estimates for all $v(t) \in \mathcal{M}$:

$$\sup_{t\in R}\|v(t)\| \le c\nu^{-1}, \quad \sup_{t\in R}\|v(t)\|_1 \le c\nu^{-1}, \tag{6.10}$$

$$\frac{1}{t-\tau}\int_\tau^t \|v(\xi)\|_2^2\mathrm{d}\xi \le c\nu^{-2}[1+\nu^{-1}(t-\tau)^{-1}], \quad -\infty < \tau < t < \infty, \tag{6.11}$$

which are better than (6.3) for large ν^{-1}.

Using these estimates we can satisfy the requirement (6.8) if N satisfies the inequality:

$$N \le c_2\nu^{-2}\left|\ln\frac{1}{\nu}\right| + c_2', \text{ with some } c_2, c_2' \in \mathbb{R}^+. \tag{6.12}$$

In the case $m=3$ (for three-dimensional Ω) we also can estimate the number of determining modes for any invariant set $\mathcal{A}$ bounded in H_1, following the same procedure. For both boundary conditions (for sticking and periodic) the estimates have the form

$$N \le c_3\nu^{-9}m_1^3 + c_3', \quad c_3, c_3' \in \mathbb{R}^+, \tag{6.13}$$

where $m_1 \equiv \sup_{v\in\mathcal{A}}\|v\|_1$.

So the following Theorem holds:

Theorem 6.1 *The number N_1 of determining modes for $\mathcal{M}$, in the case $m = 2$ and non-slip boundary conditions, has the majorant indicated in* (6.9) *and for periodic boundary conditions the one in* (6.12). *In the case $m = 3$ with either sticking or periodic conditions the majorants of number N_1 for invariant sets $\mathcal{A}$ bounded in H_1 have the form* (6.13).

Majorants of fractal dimensions

Let us denote by $d_f^{(s)}(\mathcal{A})$ the fractal dimension of a compact set $\mathcal{A}$ as a subset of space H_s. Generally the finiteness of $d_f^{(s+\epsilon)}(\mathcal{A})$, $\epsilon > 0$, for a compact $\mathcal{A}$ in the space $H_{s+\epsilon}$, does not follow from the finiteness of $d_f^{(s)}(\mathcal{A})$. But for the Navier–Stokes equations and for some other partial differential equations it is possible to evaluate $d_f^{(s)}(\mathcal{A})$ for any s using theorems from Chapter 4. For the Navier–Stokes equations we have computed the majorants for $d_f^{(k)}(\mathcal{M})$, $k = 0, 1$, in the case $m = 2$ and for $d_f^{(k)}(\mathcal{A})$, $k = 0$, 1, in the case $m = 3$. Namely

Theorem 6.2 *The numbers $d_f^{(k)}(\mathcal{M})$, $k=0$, 1, for the two-dimensional Navier–Stokes equations with sticking boundary conditions, have the following majorants:*

$$d_f^{(0)}(\mathcal{M}) \leq c_4 \nu^{-4} + c_4', \tag{6.14}$$

$$d_f^{(1)}(\mathcal{M}) \leq c_5 \nu^{-2} m_1^2 \left[\left(\ln \frac{1}{\nu} \right)^2 + (\ln m_1)^{1/2} \right] + c_5', \tag{6.15}$$

where $m_1 = \sup_{v \in \mathcal{M}} \|v\|_1$. For periodic boundary conditions, we have

$$d_f^{(k)}(\mathcal{M}) \leq c_6 \nu^{-2} \left| \ln \frac{1}{\nu} \right| + c_6', \quad k = 0,\ 1. \tag{6.16}$$

The majorants of $d_f^{(k)}(\mathcal{A})$ for the three-dimensional Navier–Stokes equations with sticking or periodic boundary conditions have the form

$$d_f^{(k)}(\mathcal{A}) \leq c_7 \nu^{-9} m_1^3 + c_7', \quad k = 0,\ 1, \tag{6.17}$$

where $m_1 = \sup_{v \in A} \|v\|_1$.

The constants c_ℓ, c_ℓ' are determined by the norm $\|h\|$ of the source term (the forces) h in the Navier–Stokes equations and in (6.15), (6.16) *by the norm $\|h\|_1$. In addition, they also depend on Ω.*

7

Evolution equations of hyperbolic type

7.1 Introduction

We now investigate semigroups originated by the problem

$$\partial_{tt}^2(t) + \nu\partial_t v(t) + Av(t) + f(v(t)) = h,$$
$$v\,|_{t=0} = \Psi_0, \quad \partial_t v\,|_{t=0} = \Psi_1. \tag{7.1}$$

Here $v = \text{constant} > 0$, A is a linear unbounded self-adjoint positively defined operator in the Hilbert space H, which has a completely continuous inverse operator; h is a fixed element of H, f is a certain nonlinear (generally unbounded) operator. Below we shall formulate conditions for f under which the problem (7.1) is globally and uniquely solvable and a semigroup it generates possesses some properties enabling us to find its minimal attractors. By using the procedure of Chapter 5, for an operator A defined in a dense domain, we introduce a space-scale $H_s(A)$, $s \in \mathbb{R}$. To apply the results of Part I, we shall reformulate the problem (7.1) as the Cauchy problem for first order equations.

Usually the vector function $\vec{v}(t) = \begin{pmatrix} v_0(t) \\ v_1(t) \end{pmatrix}$ is chosen to be unknown, where v_0 is the solution $v(t)$ of the problem (7.1) and $v_1(t) = \partial_t v(t)$.

In this variant the problem (7.1) acquires the form

$$\begin{aligned} &\partial_t v_0(t) = v_1(t), \\ &\partial_t v_1(t) = -Av_0(t) - \nu v_1(t) - f(v_0(t)) + h, \\ &v_0\,|_{t=0} = \Psi_0, \quad v_1\,|_{t=0} = \Psi_1 \end{aligned} \tag{7.2}$$

or, what is the same,

$$\begin{aligned} &\partial_t \vec{v}(t) = a\vec{v}(t) + \vec{f}(\vec{v}(t)) + \vec{h}; \\ &\vec{v}\big|_{t=0} = \vec{\Psi}, \end{aligned} \tag{7.3}$$

where

$$a = \begin{pmatrix} 0 & I \\ -A & -\nu I \end{pmatrix}, \quad \vec{f}(\vec{v}(t)) = \begin{pmatrix} 0 \\ -f(v_0(t)) \end{pmatrix},$$
$$\vec{h} = \begin{pmatrix} 0 \\ h \end{pmatrix}, \; \vec{\Psi} = \begin{pmatrix} \Psi_0 \\ \Psi_1 \end{pmatrix}.$$

However, for establishing some important properties of a semigroup generated by the problem (7.1) and associated problems, we use another formulation. That is we introduce the vector-function $\vec{v}(t, \alpha)$ defined by $\vec{v}(t)$ with the help of the equality

$$\vec{v}(t, \alpha) = C(\alpha)\vec{v}(t), \quad C(\alpha) = \begin{pmatrix} 1 & 0 \\ \alpha & 1 \end{pmatrix},$$

where α is a positive number subjected to certain restrictions. The vector function $\vec{v}(t, \alpha)$ is connected with the solution $v(t)$ of (7.1) by the equalities:

$$v_0(t, \alpha) = v(t), \quad v_1(t, \alpha) = \alpha v(t) + \partial_t v(t).$$

It is easy to prove that, if $\vec{v}(t)$ is a solution of (7.3), then $\vec{v}(t, \alpha)$ is the solution of the problem

$$\begin{aligned} \partial_t \vec{v}(t, \alpha) &= a(\alpha)\vec{v}(t, \alpha) + \vec{f}(\vec{v}(t, \alpha)) + \vec{h}, \\ \vec{v}|_{t=0} &= \vec{\Psi}(\alpha), \end{aligned} \tag{7.4}$$

where

$$a(v) = \begin{pmatrix} -\alpha I & I \\ -A(\alpha) & -(\nu - \alpha)I \end{pmatrix},$$

$$A(\alpha) = A - \alpha(\nu - \alpha)I, \; \vec{\Psi}(\alpha) = \begin{pmatrix} \Psi_0 \\ \alpha\Psi_0 + \Psi_1 \end{pmatrix};$$

and vice versa, if $\vec{v}(t, \alpha)$ is a solution of the problem (7.4), the function $\vec{v}(t) = C^{-1}(\alpha)\vec{v}(t,\alpha) = C(-\alpha)\vec{v}(t, \alpha)$ is a solution of the problem (7.3).

In terms of the components $v_0(t, \alpha)$, $v_1(t, \alpha)$ of $\vec{v}(t, \alpha)$ the system (7.4) looks like:

$$\begin{aligned} \partial_t v_0(t, \alpha) &= -\alpha v_0(t, \alpha) + v_1(t, \alpha), \\ \partial_t v_1(t, \alpha) &= -A(\alpha)v_0(t, \alpha) - (\nu - \alpha)v_1(t, \alpha) - f(v_0(t, \alpha)) + h, \\ v_0|_{t=0} = \Psi_0 &\equiv \Psi_0(\alpha), \quad v_1|_{t=0} = \alpha\Psi_0 + \Psi_1 \equiv \Psi_1(\alpha). \end{aligned} \tag{7.5}$$

It is clear that for $\alpha = 0$, the problems (7.4) and (7.3) coincide and $a(0) = a$, $\vec{\Psi}(0) = \vec{\Psi}$.

The parameter α will satisfy the following conditions

$$\alpha \in \left[0, \frac{\nu}{2}\right], \quad \frac{\lambda_1(A(\alpha))}{\lambda_1(A)} = 1 - \frac{\alpha(\nu-\alpha)}{\lambda_1(A)} \equiv m^2(\alpha) > 0. \tag{7.6}$$

Here $\lambda_1(A(\alpha)) \equiv \lambda_1(\alpha)$ is the first eigenvalue of the operator $A(\alpha)$. As before, we shall denote the complete orthonormalized system of the eigenelements of the operator A by $\{\phi_k\}_{k=1}^{\infty}$ and the corresponding eigenvalues by $\{\lambda_k\}_{k=1}^{\infty}$. The eigenvalues of the operator $A(\alpha)$ are equal to $\lambda_k(A(\alpha)) = \lambda_k(A) - \alpha(\nu-\alpha) \equiv \lambda_k(\alpha)$ and the eigenfunctions of $A(\alpha)$ and A are the same. We shall use the space-scales $H_s(A(\alpha)) \equiv H_{s,\,\alpha}$, $s \in \mathbb{R}$, constructed with the help of the operator $A(\alpha)$ as described in Chapter 5. The scalar product in $H_{s,\,\alpha}$ will be denoted by the symbol $(\cdot,\cdot)_{s,\,\alpha}$ and the norm by $\|\cdot\|_{s,\,\alpha}$. When $\alpha = 0$, instead of $H_{s,\,0}(\cdot,\cdot)_{s,\,0}$ and $\|\cdot\|_{s,\,0}$ we shall use H_s, $(\cdot,\cdot)_s$ and $\|\cdot\|_s$ (in accordance with Chapter 6).

Under the conditions (7.6), $H_{s,\,\alpha}$ and H_s coincide as sets and their norms are equivalent. In fact, elementary calculations give the following inequalities

$$\begin{aligned} m^s(\alpha)\|u\|_s \le \|u\|_{s,\,\alpha} &\equiv \left\|A^{s/2}(\alpha)u\right\| \\ &\le \left\|A^{s/2}u\right\| \equiv \|u\|_s \quad \text{for } s \ge 0, \\ \|u\|_s \le \|u\|_{s,\,\alpha} &\le m^s(\alpha)\|u\|_s \quad \text{for } s \le 0, \end{aligned} \tag{7.7}$$

where $\|\cdot\|$, as usual, is the norm in H.

As the phase space for problem (7.4) we shall take the space $X_{s,\,\alpha} \equiv H_{s+1,\,\alpha} \times H_{s,\,\alpha}$. The scalar product in $X_{s,\,\alpha}$ is defined by

$$\begin{aligned} (u, v)_{X_{s,\alpha}} &= (u_0, v_0)_{s+1,\alpha} + (u_1, v_1)_{s,\alpha} \\ &\equiv \left(A^{(s+1)/2}(\alpha)u_0, A^{(s+1)/2}(\alpha)v_0\right) + \left(A^{s/2}(\alpha)u_1, A^{s/2}(\alpha)v_1\right), \end{aligned} \tag{7.8}$$

where $(\cdot,\cdot)$, as usual, is the scalar product in H. The norm in $X_{s,\alpha}$ will be denoted by the symbol $\|\cdot\|_{X_{s,\,\alpha}}$; when $\alpha = 0$ we shall use the notation X_s, $(\cdot,\cdot)_{X_s}$ and $\|\cdot\|_{X_s}$.

From (7.7) it follows that for an arbitrary $\vec{u} \in X_s$ (or, what is the same, $\vec{u} \in X_{s,\,\alpha}$)

$$\begin{aligned} m^{s+1}(\alpha)\|\vec{u}\|_{X_s} \le \|\vec{u}\|_{X_{s,\alpha}} &\le \|\vec{u}\|_{X_s}, \quad s \ge 0, \\ m^{s+1}(\alpha)\|\vec{u}\|_{X_s} \le \|\vec{u}\|_{X_{s,\alpha}} &\le m^s(\alpha)\|\vec{u}\|_{X_s}, \quad s \in [-1, 0], \\ \|\vec{u}\|_{X_s} \le \|\vec{u}\|_{X_{s,\alpha}} &\le m^s(\alpha)\|\vec{u}\|_{X_s}, \quad s \le -1, \end{aligned} \tag{7.9}$$

The same space-scale, $X_{s,\alpha}$, $s \in \mathbb{R}$, is constructed from the space $X_{0,\alpha} \equiv H_1(A(\alpha)) \times H$ by the standard procedure described in Chapter 5, with the help of the unbounded operator $\vec{A}(\alpha) \equiv \begin{pmatrix} A(\alpha) & 0 \\ 0 & A(\alpha) \end{pmatrix}$:

$$\mathcal{D}(\vec{A}(\alpha)) \equiv \mathcal{D}\big(A^{3/2}(\alpha)\big) \times \mathcal{D}(A(\alpha)) \subset X_0 \to X_0.$$

The latter enjoys all the properties required by this procedure. The spectrum of $\vec{A}(\alpha)$ consists of the numbers $\{\lambda_k^{\pm} = \lambda_k(\alpha)\}_{k=1}^{\infty}$, each $\lambda_k(\alpha)$ corresponding to two linearly independent eigenvectors, normalized in X_0:

$$\vec{\Psi}_k^{+}(\alpha) = \begin{pmatrix} \lambda_k^{-1/2}(\alpha)\phi_k \\ 0 \end{pmatrix} \quad \text{and} \quad \vec{\Psi}_k^{-}(\alpha) = \begin{pmatrix} 0 \\ \phi_k \end{pmatrix}.$$

The operator $\vec{A}^{1/2}(\alpha)$ establishes a one-to-one correspondence between $X_{s+1,\alpha}$ and $X_{s,\alpha}$, and $\left\|\vec{A}^{1/2}\vec{u}\right\|_{X_{s,\alpha}} = \|\vec{u}\|_{X_{s+1,\alpha}}$.

Let us introduce one more notation,

$$a_0(\alpha) \equiv \begin{pmatrix} 0 & I \\ -A(\alpha) & 0 \end{pmatrix},$$

which is the principal part of the operator $a(\alpha)$. It is easy to verify that $a_0(\alpha)$ also gives a one-to-one correspondence between $X_{s+1,\alpha}$ and $X_{s,\alpha}$, and $\|a_0(\alpha)\vec{u}\|_{X_{s,\alpha}} = \|\vec{u}\|_{X_{x+1,\alpha}}$. Moreover

$$(a_0(\alpha)\vec{u}, \vec{v})_{X_{s,\alpha}} = -(\vec{u}, a_0(\alpha)\vec{v})_{X_{s,\alpha}} \tag{7.10}$$

for arbitrary $\vec{u}, \vec{v} \in X_{s+1,\alpha}$.

We begin our analysis of the problem (7.1), with the investigation of some linear problems.

7.2 Linear problems

The problem

$$\partial_t \vec{u}(t) = a\vec{u}(t) + \vec{g}(t), \quad \vec{u}\,|_{t=0} = \vec{\Psi} \tag{7.11}$$

is the object of our attention. Here $a = \begin{pmatrix} 0 & I \\ -A & -\nu I \end{pmatrix}$ is the same as in the preceding section, $\vec{g}(\cdot) = \begin{pmatrix} 0 \\ g(\cdot) \end{pmatrix}$ is a fixed element of $L_{p,loc}(\mathbb{R}, X_s)$ with $p \geq 1$ and $s \in \mathbb{R}$, and $\vec{u}(t) = \begin{pmatrix} u_0(t) \\ u_1(t) \end{pmatrix}$ is function we seek. We want to prove the solvability of (7.11) in the space X_s for an arbitrary $\vec{\Psi} \in X_s$. The main energy relation for the problem is

$$\frac{1}{2}\frac{d}{dt}\left\|\vec{u}(t)\right\|_{X_s}^2 = -\nu\left\|u_1(t)\right\|_s{}^2 + (g(t), u_1(t))_s, \tag{7.12}$$

which follows from (7.11) multiplied in X_s by $\vec{u}(t)$ and from the property (7.10) when $\alpha = 0$. By using (7.12) it is easy to estimate $\|u(t)\|_{X_s}$ through $\|\vec{\Psi}\|_{X_s}$ and $\left|\int_0^t \|g(\tau)\|_s d\tau\right|$. But it does not reflect the important property of the solutions of the problem (7.11): that is, the exponential decay of $\|\vec{u}(t)\|_{X_s}$, when $t \to +\infty$, of solutions of the homogeneous equation (7.11) (i.e. when $g(t) \equiv 0$) and the boundedness of $\left\|\vec{u}(t)\right\|_{X_s}$ on the semi-axis $t \in \mathbb{R}^+$ in the general case if $\sup_{t\in\mathbb{R}}\|g(t)\|_s < +\infty$. This can be proved by different means. One of them is to develop the solution $\vec{u}(t)$ in a Fourier series by the eigenelements of the operator a. The expansion enables one to know the behaviour of $\|\vec{u}(t)\|_{X_s}$ when $t \to \infty$. However we would prefer a different procedure, keeping in mind the further applications of the results of Part I to the problem (7.3). As in section 7.1, we introduce the functions

$$\vec{u}(t, \alpha) \equiv C(\alpha)\vec{u}(t),\ C(\alpha) = \begin{pmatrix} 1 & 0 \\ \alpha & 1 \end{pmatrix}, \tag{7.13}$$

where $\vec{u}(t)$ are solutions of (7.11) and α is a number obeying the inequalities (7.6). The $\vec{u}(t, \alpha)$ are solutions of the following problem:

$$\begin{aligned} \partial_t\vec{u}(t, \alpha) &= a(\alpha)\vec{u}(t, \alpha) + \vec{g}(t), \\ \vec{u}\,|_{t=0} &= \vec{\Psi}(\alpha) = \begin{pmatrix} \Psi_0 \\ \alpha\Psi_0 + \Psi_1 \end{pmatrix} \end{aligned} \tag{7.14}$$

where $a(\alpha) = \begin{pmatrix} -\alpha I & I \\ -A(\alpha) & -(\nu-\alpha)I \end{pmatrix}$. In coordinates (7.14) looks like

$$\begin{aligned} \partial_t u_0(t, \alpha) &= -\alpha u_0(t, \alpha) + u_1(t, \alpha), \\ \partial_t u_1(t, \alpha) &= -A(\alpha)u_0(t, \alpha) - (\nu - \alpha)u_1(t, \alpha) + g(t), \\ u_0\,|_{t=0} &= \Psi_0(\alpha) = \Psi_0, u_1\,|_{t=0} = \Psi_1(\alpha) = \alpha\Psi_0 + \Psi_1. \end{aligned} \tag{7.15}$$

The following relation holds:

$$\begin{aligned} \frac{1}{2}\frac{\mathrm{d}}{\mathrm{d}t}\|u(t, \alpha)\|_{X_{s,\alpha}}^2 &= -\alpha\|u_0(t, \alpha)\|_{s+1,\alpha}^2 - (\nu - \alpha)\|u_1(t, \alpha)\|_{s,\alpha}^2 \\ &\quad + (g(t), u_1(t, \alpha))_{s,\alpha} \end{aligned} \tag{7.16}$$

which is the analogue of (7.12) for (7.14). It is the result of the multiplication of (7.14) in $X_{s,\alpha}$ by $\vec{u}(t, \alpha)$, bearing in mind property (7.10). The inequalities

$$\begin{aligned} -\nu\|\vec{u}(t, \alpha)\|_{X_{s,\alpha}} - \|\vec{g}(t)\|_{s,\alpha} &\le \frac{\mathrm{d}}{\mathrm{d}t}\|\vec{u}(t, \alpha)\|_{X_{s,\alpha}} \\ &\le -\alpha\|\vec{u}(t, \alpha)\|_{X_{s,\alpha}} + \|\vec{g}(t)\|_{s,\alpha} \end{aligned} \tag{7.17}$$

follow from (7.16), because we assumed that $\alpha \in [0, \nu/2]$, while from (7.17) we may deduce the following estimates

$$\|\vec{u}(t,\alpha)\|_{X_{s,\alpha}} \le e^{-\alpha t}\|\vec{u}(0,\alpha)\|_{X_{s,\alpha}} + \int_0^t e^{-\alpha(t-\tau)}\|g(\tau)\|_{s,\alpha}\,d\tau, \quad t \in \mathbb{R}^+, \tag{7.18$_1$}$$

$$\|\vec{u}(t,\alpha)\|_{X_{s,\alpha}} \le e^{-\nu t}\|\vec{u}(0,\alpha)\|_{X_{s,\alpha}} + \int_t^o e^{-\nu(t-\tau)}\|g(\tau)\|_{s,\alpha}\,d\tau, \quad t \in \mathbb{R}^-. \tag{7.18$_2$}$$

These inequalities allow us to obtain estimates of the same kind for $\|\vec{u}(t)\|_{X_s}$. In fact, for an arbitrary $u \in X_s$ and $\vec{u}(\alpha) \equiv C(\alpha)\vec{u}$, where $C(\alpha)$ is defined in (7.13), the inequalities

$$\|\vec{u}(\alpha)\|_{X_s} \le \mu_1(\alpha)\|\vec{u}\|_{X_s} \le \mu_1^2(\alpha)\|\vec{u}(\alpha)\|_{X_s}, \tag{7.19$_1$}$$

where $\mu_1(\alpha) = \max\left\{\sqrt{1+\alpha}, \sqrt{1+\alpha\lambda_1^{-1}+\alpha^2\lambda_1^{-1}}\right\}$ are confirmed routinely. From (7.19$_1$) and (7.9) we deduce the inequalities

$$\begin{aligned} \|\vec{u}\|_{X_s} &\le m_1(s,\alpha)\|\vec{u}(\alpha)\|_{X_{s,\alpha}}, \\ \|\vec{u}(\alpha)\|_{X_{s,\alpha}} &\le m_2(s,\alpha)\|\vec{u}\|_{X_s}, \end{aligned} \tag{7.19$_2$}$$

where

$$m_1(s,\alpha) = \begin{cases} \mu_1(\alpha)m^{-(s+1)}(\alpha), & \text{for } s \ge -1 \\ \mu_1(\alpha), & \text{for } s \le -1 \end{cases}$$

$$m_2(s,\alpha) = \begin{cases} \mu_1(\alpha), & \text{for } s \ge 0 \\ \mu_1(\alpha)m^s(\alpha), & \text{for } s \le 0 \end{cases}$$

For α satisfying requirements (7.6), because of (7.19$_2$) and (7.7) we obtain from (7.18$_k$) the main a priori estimates:

$$\|\vec{u}(t)\|_{X_s} \le m_3(s,\alpha)e^{-\alpha t}\|\vec{u}(0)\|_{X_s} + m_4(s,\alpha)\int_0^t e^{-\alpha(t-\tau)}\|g(\tau)\|_s\,d\tau, \tag{7.20$_1$}$$

$$\|\vec{u}(t)\|_{X_s} \le m_3(s,\alpha)e^{-\nu t}\|\vec{u}(0)\|_{X_s} + m_4(s,\alpha)\int_t^0 e^{-\nu(t-\tau)}\|g(\tau)\|_s d\tau, \tag{7.20$_2$}$$

for $t \in \mathbb{R}^+$ and $t \in \mathbb{R}^-$ respectively, where

$$m_3(\mathrm{s},\alpha) = m_1(s,\alpha)m_2(\mathrm{s},\alpha)$$

and

$$m_4(s,\ \alpha) = m_1(s,\alpha) \qquad \text{if } s \geq 0,$$
$$m_4(s,\ \alpha) = m_1(s,\alpha)m^s(\alpha) \quad \text{if } s \leq 0$$

for solutions of the problem (7.11). The explicit dependence of $m_k(\cdot)$ on s and on α is not important for us. The essential things are:

(a) the requirements (7.6) enable us to choose α positive and deduce that, for an arbitrary $s \in \mathbb{R}$, $\|\vec{u}(t)\|_{X_s}$ tends to zero when $t \to +\infty$.
(b) when α tends to zero, $\mu_1(\alpha)$, $m(\alpha)$ and $m_k(s, \alpha)$ tend to 1.

Now let us go on to prove the unique solvability of the problem (7.11).

We shall call a *weak solution* of the problem (7.11) a function $\vec{u} : \mathbb{R} \to X_r$ for which all projections $(u_0(t), \phi_k)$, $(u_1(t), \phi_k)$, $k = 1, 2, \ldots$ are absolutely continuous functions of $t \in \mathbb{R}$ and, for almost any t, satisfy the following equalities:

$$\begin{aligned}
&\partial_t(u_0(t), \phi_k) = (u_1(t), \phi_k);\\
&\partial_t(u_1(t), \phi_k) = -\lambda_k(u_0(t), \phi_k) - \nu(u_1(t), \phi_k) + (g(t), \phi_k);\\
&(u_0, \phi_k)\,|_{t=0} = (\Phi_0, \phi_k),\\
&(u_1, \phi_k)\,|_{t=0} = (\Phi_1, \phi_k), \ \ k = 1, 2, \ldots.
\end{aligned} \tag{7.21}$$

In this terminology the number r is deliberately ignored despite the fact it appears in the condition $\vec{u}\colon \mathbb{R} \to X_r$. Its value is insignificant, only the fact that $r > -\infty$ being important. It should be remembered that for the elements $u \in H_s$ and $v \in H_{-s}$, (u, v) is the number $(A^{s/2}u, A^{-s/2}v)$ — the inner product in H of elements $A^{s/2}u \in H$ and $A^{-s/2}v \in H$. This is the meaning of the brackets $(u_\ell(t), \phi_k)$ in (7.21).

It is easy to prove the following theorem

Theorem 7.1 *Problem* (7.11) *has no more than one weak solution.*

We shall apply this theorem to linear and nonlinear problems as a means of identifying their solutions when we have preliminary and incomplete information, but when we know that they are weak solutions of a problem of the type (7.11).

Let us prove the following existence theorem:

Theorem 7.2 *If $g \in L_{p,loc}(\mathbb{R}, H_s)$, $s \in \mathbb{R}$, $p \geq 1$ then problem* (7.11) *for an arbitrary $\vec{\Psi} \in X_s$ has the unique solution $\vec{u} \in C(\mathbb{R}, X_s)$ with $\partial_t\vec{u} \in L_{p,loc}(\mathbb{R}, X_{s-1})$. For almost all t it satisfies the equation* (7.11) *(in space X_{s-1}) and the energy relation* (7.12). *For this solution the estimates*

(7.20_k) are true (for all t). Any weak solution of problem (7.11) *coincides with this $\vec{u}$. If, additionally, $g \in C(\mathbb{R}, H_{s-1})$ then $\partial_t \vec{u} \in C(\mathbb{R}, X_{s-1})$ and equation (7.11) is fulfilled for all $t \in \mathbb{R}$.*

It is easy to prove that all the statements of Theorem 7.2 are true if $\Psi_0 = \sum_{k=1}^N a_k \phi_k$, $\Psi_1 = \sum_{k=1}^N b_k \phi_k$, and $g(t) = \sum_{k=1}^N g_k(t)\phi_k$ with $g_k \in L_{p,\,loc}(\mathbb{R})$ or $g_k \in C(\mathbb{R})$.

Now let $\vec{\Psi}$ be an arbitrary element of X_s and $g(\cdot)$ an arbitrary element of $L_{p,loc}(\mathbb{R}, H_s)$. The sums of their Fourier-series, $\vec{\Psi}^N$ and $g^N(\cdot)$, containing only $\phi_1, \dots, \phi_N$, approximate $\vec{\Psi}$ and $g(\cdot)$ when $N \to \infty$ as follows: $\vec{\Psi}^n \to \vec{\Psi}$ in the norm X_s, and $g^N(\cdot) \to g(\cdot)$, in the norm $L_p((-T, T), H_s)$, for an arbitrary $T < +\infty$. Because of this and the fact that the estimates (7.20_k) hold for the solution $\vec{u}^N(t)$ of the problem (7.11), corresponding to $\vec{\Psi}^N$ and $g^N(\cdot)$, and their differences $\vec{u}^{N_1}(t) - \vec{u}^{N_2}(t)$ for arbitrary $N_1, N_2 < \infty$, the sequence $\{\vec{u}^N(\cdot)\}_{N=1}^\infty$ is fundamental in the spaces $C([-T, T], X_s)$. for any $T < +\infty$. Since the spaces $C([-T, T], X_s)$ are complete, the sequence $\{\vec{u}^N(\cdot)\}_{N=1}^\infty$ converges in their norms to an element $\vec{u} \in C(R, X_s)$. Hence $a\vec{u}^N(\cdot)$ converges to $a\vec{u}(\cdot)$ in the norm of $C([-T, T], X_{s-1})$ (one must remember that $a \in L(X_s \to X_{s-1})$). Furthermore, as $\partial_t \vec{u}^N(t) = a\vec{u}^N(t) + \vec{g}^N(t)$ and $\vec{g}^N(\cdot) \to \vec{g}(\cdot)$ in the norms of $L_p([-T, T], X_s)$, $T < +\infty$, then $\partial_t \vec{u}^N(\cdot)$ converges in the norms of $L_p([-T, T], X_{s-1})$ to an element of $L_{p,loc}(\mathbb{R}, X_{s-1})$, which, according to the theory of generalized derivatives, is equal to $\partial_t \vec{u}(\cdot)$. If, additionally, $g \in C(\mathbb{R}, H_{s-1})$, then $\vec{g}^N(\cdot)$ converges to $\vec{g}(\cdot)$ in the norms of $C([-T, T], X_{s-1})$ and $\partial_t \vec{u}^N(\cdot)$ converges to $\partial_t \vec{u}$ in the same norms. In this case equation (7.11) will be fulfilled (in X_{s-1}) for all t, and in the first case (i.e. for $g \in L_{p,loc}(\mathbb{R}, X_s)$) for almost all t. Relation (7.12) is fulfilled (for almost all t), since it is true for all $u^N(t)$ with $g(t) = g^N(t)$, and each term of its right hand side converges to the corresponding term of the right hand side of (7.12) in the norms of $L_p((-T, T))$, $T < +\infty$.

The uniqueness follows from Theorem 7.1 since the solution $\vec{u}(t)$ is a weak solution of problem (7.11).

Theorem 7.2 for the case $g(t) \equiv 0$ guarantees the existence of the operators $U_t \in L(X_s \to X_s)$ associating to $\vec{\Psi} \in X_s$ the solution $\vec{u}(t)$ of the problem (7.11).

$$\partial_t \vec{u}(t) = a\vec{u}(t), \; \vec{u}\,|_{t=0} = \vec{\Psi}. \tag{7.22}$$

The inequalities (7.20_k) give the following estimates:

$$\|U_t\|_{L(X_s \to X_s)} \le m_3(\alpha, s)e^{-\alpha t}, \quad t \in \mathbb{R}^+, \tag{7.23$_1$}$$

$$\|U_t\|_{L(X_s \to X_s)} \le m_3(\alpha, s)e^{-\nu t}, \quad t \in \mathbb{R}^-. \tag{7.23$_2$}$$

It is obvious that $\{U_t,\ t \in \mathbb{R},\ X_s\}$ is a continuous group of linear bounded operators.

As in the finite-dimensional case, the solution $\vec{u}(t)$ of problem (7.11) may be represented by means of Duhamel's principle as follows

$$\vec{u}(t) = U_t\vec{\Psi} + \int_0^t U_{t-\tau}\vec{g}(\tau)d\tau. \tag{7.24}$$

This can be proved using the above mentioned properties of the operators U_t and the fact that $a \in L(X_s \to X_{s-1})$. But, otherwise, the representation (7.24) is true for the finite-dimensional approximations $\vec{u}^N(t)$ and the estimates (7.20_k) allow us to pass to the limit when $N \to \infty$ and to obtain (7.24) for $\vec{u}(t)$ as the limit of $\vec{u}^N(t)$.

Problem (7.11), as proved above, is connected with the problem:

$$\begin{aligned} &\partial_{tt}^2 u(t) + \nu\partial_t u(t) + Au(t) = g(t), \\ &u\mid_{t=0} = \Psi_0, \quad \partial_t u\mid_{t=0} = \Psi_1 \end{aligned} \tag{7.25}$$

in the following way. The first component $u_0(t)$ of the solution $\vec{u}(t)$ is the solution $u(t)$ of problem (7.25) and the second component of $\vec{u}(t)$ is equal to $\partial_t u(t)$. Therefore from Theorem 7.2 for problem (7.25) we get:

Theorem 7.2′ *If $g \in L_{p,loc}(\mathbb{R}, H_s)$, $s \in \mathbb{R}$, $p \geq 1$, then problem* (7.25) *has the unique solution $u \in C(\mathbb{R}, H_{s+1})$ with $\partial_t u \in C(\mathbb{R}, H_s)$ and $\partial_t^2 u \in L_{p,loc}(\mathbb{R}, H_{s-1})$. The solution $u(t)$ for almost all $t \in \mathbb{R}$ satisfies the equality* (7.25) *(in H_{s-1}) and the energy identity*

$$\frac{1}{2}\frac{\mathrm{d}}{\mathrm{d}t}(\|u(t)\|_{s+1}^2 + \|\partial_t u(t)\|_s^2) = -\nu\ \|\partial_t u(t)\|_s^2 + (g(t),\ \partial_t u(t))_s. \tag{7.26}$$

If, in addition, $g \in C(\mathbb{R}, H_{s-1})$ then equation (7.25) *is satisfied for each $t \in \mathbb{R}$ (in H_{s-1}). The estimates* (7.20_k) *are fulfilled for*

$$\|\vec{u}(t)\|_{X_s} := \left[\|u(t)\|_{s+1}^2 + \|\partial_t u(t)\|_s^2\right]^{1/2}.$$

The estimate

$$\|\partial_{tt}^2 u(t)\|_{s-1} \leq \|u(t)\|_{s+1} + \nu\|\partial_t u(t)\|_{s-1} + \|g(t)\|_{s-1} \tag{7.27}$$

is also true.

The last statement follows from (7.25). The following theorem holds:

Theorem 7.3 *If the function $g\colon \mathbb{R} \to H_r$ is absolutely continuous and $\partial_t g \in L_{p,loc}(\mathbb{R}, H_r)$, $r \in \mathbb{R}$, $p \geq 1$, then for the solution $\vec{u} = \binom{u_0}{u_1}$ of problem* (7.11) *with $\vec{\Psi} = 0$, $\partial_t\vec{u} \in C(\mathbb{R}, X_r)$, $u_0 \in C(\mathbb{R}, H_{r+2})$ and for $t \in \mathbb{R}^+$ the estimates*

$$\|\partial_t \vec{u}(t)\|_{X_r} \leq m_3(\alpha, r)e^{-\alpha t}\|g(0)\|_r + m_4(\alpha, r)\int_0^t e^{-\alpha(t-\tau)}\|\partial_t g(\tau)\|_r\, d\tau, \tag{7.28}$$

$$\|u_0(t)\|_{r+2} \leq c\|\partial_t \vec{u}(t)\|_{X_r} + \|g(t)\|_r, \tag{7.29}$$

hold. Here $c = \sqrt{2}\max\{1, \nu\lambda_1^{-1/2}\}$.

The proof is based on Theorems 7.1 and 7.2. Theorem 7.2 guarantees the existence of $\vec{u} \in C(\mathbb{R}, X_r)$, the estimates (7.20_k) with $\vec{u}(0) = 0$ and $s = r$, as well as $\partial_t \vec{u} \in C(\mathbb{R}, X_{r-1})$.

The latter implies $a\partial_t \vec{u} = \partial_t a\vec{u} \in C(\mathbb{R}, X_{r-2})$ and from $\vec{u} \in C(\mathbb{R}, X_r)$ it follows that $a\vec{u} \in C(\mathbb{R}, X_{r-1})$. So, all the members in the equation (7.11) are elements of $C(\mathbb{R}, X_{r-1})$. Furthermore, since both terms on the right hand side of (7.11) are differentiable in t and their derivatives are elements of $C(\mathbb{R}, X_{r-2})$ and $L_{p,loc}(\mathbb{R}, X_r)$, respectively, then $\partial_t \vec{u}$ also is differentiable in t, $\partial_{tt}\vec{u} \in L_{p,loc}(\mathbb{R}, X_{r-2})$ and $\partial_t \vec{u}$ is a solution of the problem

$$\partial_t \vec{w}(t) = a\vec{w}(t) + \partial_t \vec{g}(t), \quad \vec{w}\,|_{t=0} = \vec{g}(0) \tag{7.30}$$

belonging to $C(\mathbb{R}, X_{r-1})$. On the other hand, the same Theorem 7.2, when applied to the problem (7.30), guarantees the existence of the solution $\vec{w} \in C(\mathbb{R}, X_r)$ as $\vec{g}(0) \in X_r$ and $\partial_t \vec{g} \in L_{p,loc}(\mathbb{R}, H_r)$. Since $\partial_t \vec{u}$ and $\vec{w}$ are both weak solutions of (7.30), they must coincide according to the uniqueness theorem, i.e. $\partial_t \vec{u} = \vec{w} \in C(\mathbb{R}, X_r)$. The estimate (7.28) is nothing but the estimate (7.20_1) (with $s = r$) for $\vec{w}$ and the estimate (7.29) follows from the fact that $\vec{u}(t) = \begin{pmatrix} u_0(t) \\ u_1(t) \end{pmatrix} = \begin{pmatrix} u(t) \\ \partial_t u(t) \end{pmatrix}$, where $u(t)$ is the solution of (7.25).

7.3 On global unique solvability of the nonlinear problem

In this section we investigate the global unique solvability of problem (7.1) or, what is the same, of problem (7.3). We shall use both of these formulations and the relations $v(t) = v_0(t)$, $\partial_t v(t) = v_1(t)$ between the solutions $v(t)$ of (7.1) and $\vec{v}(t) = \begin{pmatrix} v_0(t) \\ v_1(t) \end{pmatrix}$ of (7.3). We choose X_0 as the phase-space and (according to this choice) impose on f the following conditions:

(a) $f\colon H_1 \to H$, $f(0) = 0$ and for all $u_1, u_2 \in H_1$

$$\|f(u_1) - f(u_2)\| \leq \Phi_1(\max\{\|u_1\|_1; \|u_2\|_1\})\|u_1 - u_2\|_1. \tag{7.31}$$

Here and below, Φ_k are some functions, $\Phi_k\colon \mathbb{R}^m \to \mathbb{R}^+$, their explicit form being insignificant; we only assume that these functions are continuous and nondecreasing when each of their arguments increases.

(b) For any $= 1, 2, \ldots$ and all $u_1, u_2 \in H_1$

$$|(f(u_1) - f(u_2), \phi_k)| \leq \Phi_2(k,\ \max\{\|u_1\|_1, \|u_2\|_1\})\|u_1 - u_2\|; \tag{7.32}$$

(c) f is a potential operator with the continuous potential $\mathcal{F}\colon H_1 \to \mathbb{R}$, $\mathcal{F}(0) = 0$ such that for every $u \in C(\mathbb{R}, H_1)$ with $\partial_t u \in C(\mathbb{R}, H)$ the function $\mathcal{F}(u(\cdot))\colon \mathbb{R} \to \mathbb{R}$ is absolutely continuous and

$$\frac{\mathrm{d}}{\mathrm{d}t}\mathcal{F}(u(t)) = (f(u(t)), \partial_t u(t)). \tag{7.33}$$

Moreover, for all $u \in H_1$, $\mathcal{F}(u)$ must satisfy the inequality

$$-\mathcal{F}(u) \leq \left(\frac{1}{2} - \nu_1\right)\|u\|_1^2 + c_1 \tag{7.34}$$

with some $\nu_1 \in \left(0, \frac{1}{2}\right]$ and $c_1 \in \mathbb{R}^+$.

From (7.33) it follows that

$$\mathcal{F}(u) = \int_0^1 (f(tu), u)\mathrm{d}t, \qquad \forall u \in H_1 \tag{7.35}$$

and from (7.35) and (7.31) the estimate

$$\begin{aligned} |\mathcal{F}(u)| &\leq \frac{1}{2}\Phi_1(\|u\|_1)\|u\|_1\|u\| \\ &\leq \frac{1}{2\sqrt{\lambda_1}}\Phi_1(\|u\|_1)\|u\|_1^2 \equiv \Phi_3(\|u\|_1) \end{aligned} \tag{7.36}$$

follows.

The following theorem holds:

Theorem 7.4 *Let $h \in H$, $\vec{\Psi} \in X_0$ and f satisfy the conditions (a)–(c). Then problem* (7.3) *has the unique solution $\vec{v}$ enjoying the following properties: $\vec{v} \in C(\mathbb{R}, X_0)$, $\partial_t\vec{v} \in C(\mathbb{R}, X_{-1})$; equation* (7.3) *is fulfilled for all $t \in \mathbb{R}$ in X_{-1}; $\vec{v}(t)$ is given by*

$$\vec{v}(t) = U_t(\vec{v}(0)) + \int_0^t U_{t-\tau}\begin{pmatrix} 0 \\ -f(v_0(\tau)) + h \end{pmatrix}\mathrm{d}\tau, \tag{7.37}$$

where $U_t \in L(X_0 \to X_0)$ are the solution operators of problem (7.22); the energy relation holds:

$$\frac{\mathrm{d}}{\mathrm{d}t}\mathcal{L}(\vec{v}(t)) = -\nu\|v_1(t)\|^2, t \in R, \tag{7.38}$$

where the function $\mathcal{L} : X_0 \to \mathbb{R}$ is defined by

$$\mathcal{L}(\vec{u}) := \frac{1}{2}\|u_0\|_1^2 + \frac{1}{2}\|u_1\|^2 + \mathcal{F}(u_0) - (h, u_0), \\ \vec{u} = \begin{pmatrix} u_0 \\ u_1 \end{pmatrix}. \tag{7.39}$$

For the solution $\vec{v}(t)$ the estimates (7.45)–(7.46$_k$) *hold.*

The solution operators V_t of problem (7.3) *form a continuous group $\{V_t, t \in \mathbb{R}, X_0\}$; the corresponding semigroup $\{V_t, t \in \mathbb{R}^+, X_0\}$ is bounded.*

First we obtain the basic a priori estimate for the solutions of problem (7.3). To this end we rewrite (7.3) in the form

$$\partial_t\vec{v}(t) = \alpha\vec{v}(t) + \vec{g}(t), \quad \vec{v}|_{t=0} = \vec{\Psi}, \tag{7.40}$$

where

$$\vec{g}(t) \equiv \vec{g}(t; \vec{\Psi}) := \begin{pmatrix} 0 \\ g(t; \Psi) \end{pmatrix} \\ = \vec{f}(\vec{v}(t)) + \vec{h} = \begin{pmatrix} 0 \\ -f(v_0(t)) \end{pmatrix} + \begin{pmatrix} 0 \\ h \end{pmatrix}. \tag{7.41}$$

The dependence of $\vec{g}$ on the functional parameter $\vec{\Psi}$ is due to the fact that the solution $\vec{v}$ of problem (7.3) depends on $\vec{\Psi}$.

Equation (7.40) has the form (7.11) and therefore the solutions $\vec{v}$ of (7.40) satisfies (7.12) with $s = 0$ (provided $\vec{v}$ is smooth enough, see above). The last term in (7.12), because of condition (c) with $v_1(t) = \partial_t v_0(t)$, is equal to

$$(g(t), v_1(t)) = (-f(v_0(t)) + h, \partial_t v_0(t)) \\ = +\frac{\mathrm{d}}{\mathrm{d}t}[-\mathcal{F}(v_0(t)) + (h, v_0(t))]. \tag{7.42}$$

Therefore, (7.38) is nothing but the relation (7.12) with $s = 0$.

From (7.38) and (7.36) it follows

$$\mathcal{L}(\vec{v}(t)) + \nu\int_0^t \|v_1(\tau)\|^2 d\tau = \mathcal{L}(\vec{v}(0)) \\ \leq \frac{1}{2}\|\Psi_1\|^2 + \frac{1}{2}\|\Psi_0\|_1^2 + \Phi_3(\|\Psi_0\|_1) + \|h\|\|\Psi_0\|. \tag{7.43}$$

On the other hand, from (7.34) we deduce

$$\begin{aligned}\mathcal{L}(\vec{v}(t)) &\geq \frac{1}{2}\|v_1(t)\|^2 + \frac{1}{2}\|v_0(t)\|_1^2 \\ &\quad - \left(\frac{1}{2} - \nu_1\right)\|v_0(t)\|_1^2 - c_1 - \|h\|\lambda_1^{-1/2}\|v_0(t)\|_1 \\ &\geq \frac{1}{2}\|v_1(t)\|^2 + \frac{\nu_1}{2}\|v_0(t)\|_1^2 - c_1 - (2\nu_1\lambda_1)^{-1}\|h\|^2.\end{aligned} \tag{7.44}$$

Now (7.43) and (7.44) give the "a priori" estimate:

$$\begin{aligned}&\|v_1(t)\|^2 + \nu_1\|v_0(t)\|_1^2 + 2\nu\int_0^t \|v_1(\tau)\|^2 d\tau \\ &\leq 2c_1 + (\nu_1\lambda_1)^{-1}\|h\|^2 + \|\Psi_1\|^2 + \|\Psi_0\|_1^2 \\ &\quad + 2\Phi_3(\|\Psi_0\|_1) + 2\|h\|\|\Psi_0\|_1\lambda_1^{-1/2} \\ &\equiv \Phi_4(\|\vec{\Psi}\|_{X_0}).\end{aligned} \tag{7.45}$$

In particular,

$$\|\vec{v}(t)\|_{X_0} \leq \Phi_5(\|\vec{\Psi}\|_{X_0}), \qquad t \in \mathbb{R}^+, \tag{7.46$_1$}$$

$$\|\vec{v}(t)\|_{X_0} \leq \Phi_6(|t|, \|\vec{\Psi}\|_{X_0}), \quad t \in \mathbb{R}^-. \tag{7.46$_2$}$$

The explicit form of the functions Φ_5 and Φ_6 is not significant to our purpose; note that $\Phi_6(|t|, \xi) \to +\infty$ when $t \to -\infty$.

We shall construct the approximate solutions $\vec{v}^m(t)$, $m = 1, 2, \ldots$ of (7.3) by means of the Galerkin–Faedo method using the eigenfunctions $\{\phi_k\}_{k=1}^{\infty}$ of the operator A as the basis in H_s. Namely, we look for $\vec{v}^m(t) = \begin{pmatrix} v^m(t) \\ \partial_t v^m(t)\end{pmatrix}$ where $v^m(t) = \sum_{\ell=1}^m c_\ell^m(t)\phi_\ell$ and the coefficients $c_\ell^m(t) = (v^m(t), \phi_\ell)$, $\ell = 1, \ldots, m$ are the solutions of the Cauchy problem for the following system of m ordinary differential equations:

$$\begin{aligned}&\partial_{tt}^2(v^m(t), \phi_k) + \nu\partial_t(v^m(t), \phi_k) \\ &\quad + \lambda_k(v^m(t), \phi_k) + (f(v^m(t)), \phi_k) = (h, \phi_k) \\ &c_k^m(0) = (\Psi_0, \phi_k), \partial_t c_k^m(0) = (\Psi_1, \phi_k), k = 1, \ldots, m.\end{aligned} \tag{7.47}$$

Let us prove that the "a priori" estimates (7.46$_k$) hold for $\vec{v}^m(t) := \begin{pmatrix} v^m(t) \\ \partial_t v^m(t)\end{pmatrix}$, and, hence, for any $T \in \mathbb{R}^+$:

$$\max_{t\in[-T,T]} \|\vec{v}^m(t)\|_{X_0} \leq \Phi_7\big(T, \|\vec{\Psi}\|_{X_0}\big), \; m = 1, 2, \ldots, \tag{7.48}$$

where $\Phi_7\big(T, \|\vec{\Psi}\|_{X_0}\big) = \max\big\{\Phi_5\big(\|\vec{\Psi}\|_{X_0}\big); \Phi_6\big(T; \|\vec{\Psi}\|_{X_0}\big)\big\}$.

We multiply the k-th equations (7.47) by $\partial_t c_k^m(t)$ and sum the results for $k = 1, \ldots, m$. It is easy to verify that this yields for $\vec{v}^m(t)$ the equality

(7.38), from which the estimates ($7.46_{k,k=1,2}$) follow. The estimates (7.48) for the solutions of (7.47) and our hypothesis on f guarantee the global unique solvability of the problem (7.47). The coefficients $c_\ell^m(t)$ are twice continuously differentiable in t and, hence, $v^m(t)$ and $\partial_t v^m(t)$ belong to $C(\mathbb{R}, X_0)$. To perform the limit along some subsequence $m_j \to +\infty$, let us estimate the quantities $|(\partial_{tt}^2 v^m(t), \phi_k)|$, using (7.32) and (7.47). Namely,

$$\begin{aligned} |(\partial_{tt}^2 v^m(t), \phi_k)| &\le \nu \|\partial_t v^m(t)\| + \lambda_k^{1/2} \|v^m(t)\|_1 \\ &\quad + \Phi_2(k, \|v^m(t)\|_1)\|v^m(t)\| + \|h\| \\ &\equiv \Phi_8(k, \|\vec{v}^m(t)\|_{X_0}). \end{aligned} \tag{7.49}$$

By (7.48) and (7.49) we may choose from $\{\vec{v}^m(t)\}_{m=1}^\infty$ a subsequence $\{\vec{v}^{m_j}(t)\}_{j=1}^\infty$ enjoying the following properties:

(1) $\{\vec{v}^{m_j}(\cdot)\}$ converges to some element $v \in C(\mathbb{R}, H)$ in the norms of spaces $C([-T, T], H)$, with any $T \in \mathbb{R}^+$;
(2) $\{v^{m_j}(\cdot)\}$ and $\{\partial_t v^{m_j}(\cdot)\}$ converge weakly in the Hilbert spaces $L_2((-T, T), H_1)$ and $L_2((-T, T), H)$, $\forall T \in \mathbb{R}^+$, respectively; hence, $v \in L_{2,loc}(\mathbb{R}, H_1)$ and $\partial_t v \in L_{2,loc}(\mathbb{R}, H)$ and for almost every t

$$\|v(t)\|_1 \le \underline{\lim}_{j\to\infty} \|v^{m_j}(t)\|_1 \quad \text{and} \quad \|\partial_t v(t)\| \le \underline{\lim}_{j\to\infty} \|\partial_t v^{m_j}(t)\|.$$

(3) For any (fixed) $k = 1, 2, \ldots$, $\left\{\partial_t c_k^{m_j}(\cdot) = (\partial_t v^{m_j}(\cdot), \phi_k)\right\}_{j=1}^\infty$ converges in $C([-T, T])$, $\forall T \in \mathbb{R}^+$, to $\partial_t c_k(\cdot) \equiv (\partial_t v(\cdot), \phi_k) \in C(R)$.

From (1) and the hypothesis (7.32) it follows that $\left(f\left(v^{m_j}(\cdot)\right), \phi_k\right)$ converges (as $j \to +\infty$) to $(f(v(\cdot)), \phi_k)$ in $C([-T, T])$, $\forall T \in \mathbb{R}^+$. Thus, for a fixed k, all terms in (7.47) except the first one have the appropriate limits in $C([-T,T])$, $\forall T \in \mathbb{R}^+$. Hence $\partial_{tt}^2(v^m(\cdot), \phi_k)$ also has a limit (in the same sense) and this limit is $\partial_{tt}^2(v^m(\cdot), \phi_k) \in C([-T, T])$, $\forall T \in \mathbb{R}^+$. Now we may conclude that the limit $v(\cdot)$ satisfies the equations:

$$\begin{gathered} \frac{d^2}{dt^2}(v(t), \phi_k) + \frac{d}{dt}(v(t), \phi_k) + \lambda_k (v(t), \phi_k) + (f(v(t)), \phi_k) = (h, \phi_k), \\ (v, \phi_k)\,|_{t=0} = (\Psi_0, \phi_k),\ \partial_t (v, \phi_k)\,|_{t=0} = (\Psi_1, \phi_k)\ , \quad k = 1, 2, \ldots, \end{gathered} \tag{7.50}$$

and, besides, $v \in C(\mathbb{R}, H) \cap L_{2,loc}(\mathbb{R}, H_1)$ and $\partial_t v \in L_{\infty,loc}(\mathbb{R}, H)$. Moreover, $g(\cdot, \vec{\Psi}) \equiv -f(v(\cdot)) + h$ is an element of $L_{\infty,loc}(\mathbb{R}, H)$ and

$$\begin{aligned}\|g(t, \vec{\Psi})\| &\le \Phi_1(\|v(t)\|_1)\|v(t)\|_1 + \|h\| \\ &\le \Phi_1(\Phi_7(T, \|\vec{\Psi}_0\|_{X_0}))\Phi_7(T, \|\vec{\Psi}_0\|_{X_0}) + \|h\| \\ &\equiv \Phi_8(T, \|\vec{\Psi}_0\|_{X_0}),\end{aligned} \tag{7.51}$$

for $t \in [= T, T]$. Hence $\vec{v}(\cdot) \equiv \begin{pmatrix} v(\cdot) \\ \partial_t v(\cdot) \end{pmatrix}$ may be interpreted as a weak solution of the linear problem (7.40) with $g \in L_{\infty,\,loc}(\mathbb{R}, H)$ and $\vec{\Psi} \in X_0$. On the other hand Theorem 7.2 guarantees for this problem the existence of a solution in $C(\mathbb{R}, X_0)$, and because of Theorem 7.1 this solution must coincide with $\vec{v}$. Using this result and, in addition the hypotheses on f and applying Theorem 7.2 we obtain all the statements of Theorem 7.4 except the last one and the uniqueness for problem (7.3). In order to prove the uniqueness suppose that the problem (7.3) has another solution $\vec{v}'(t)$ possessing all the properties of $\vec{v}(t)$. The difference $\vec{w}(t) = \vec{v}(t) - \vec{v}'(t)$ may be regarded as a solution of the linear problem (7.40) with the free term $\vec{g}(t) = \begin{pmatrix} 0 \\ -f(v_0(t)) + f(v_0'(t)) \end{pmatrix}$ and $\vec{\Psi} \equiv 0$. Since $\vec{w} \in C(\mathbb{R}, X_0)$ and $\vec{g}(\cdot) \in L_{\infty,loc}(\mathbb{R}, X_0)$, for $\vec{w}$ the energy equality (7.12) holds. We estimate the last term in (7.12) using (7.31) and (7.46$_k$):

$$\begin{aligned}&|(-f(v_0(t)) - f(v_0'(t)), w_1(t))| \\ &\quad\le \Phi_1(\max\{\|v_0(t)\|_1, \|v_0'(t)\|_1\})\|w_0(t)\|_1\|w_1(t)\| \\ &\quad\le \Phi_9(|t|)\|\vec{w}(t)\|^2_{X_0}.\end{aligned} \tag{7.52}$$

Hence $\vec{w}(t)$ satisfies the inequality

$$\frac{1}{2}\frac{\mathrm{d}}{\mathrm{d}t}\|\vec{w}(t)\|^2_{X_0} + \nu\|w_1(t)\|^2 \le \Phi_9(|t|)\|\vec{w}(t)\|^2_{X_0} \tag{7.53}$$

and since $\vec{w}(0) = 0$ we conclude that $\vec{w}(t) \equiv 0$. Consequently, problem (7.3) has the unique solution with the properties indicated in Theorem 7.4.

Thus we have proved that the solution operators $V_t : \vec{\Psi} \in X_0 \to \vec{v}(t) \in X_0$ are defined for all $\vec{\Psi} \in X_0$ and $t \in \mathbb{R}$; they are single-valued and because of (7.45), (7.46$_k$) are bounded (i.e. they map bounded sets into bounded sets). The family $\{V_t, t \in \mathbb{R}, X_0\}$ has the group property: $V_{t_1+t_2} = V_{t_1}V_{t_2}$ for all $t_1, t_2 \in \mathbb{R}$. In order to prove that the operators V_t are continuous on X_0, let $\vec{v}(t) = V_t(\vec{\Psi})$ and $\vec{\tilde{v}}(t) = V_t(\vec{\tilde{\Psi}})$. As in the proof of uniqueness, we have for $\vec{w}(t) = \vec{v}(t) - \vec{\tilde{v}}(t)$ the inequality (7.53) with

$$\Phi_9(|t|) = \frac{1}{2}\Phi_1(\max\{\|v_0(t)\|_1, \|\tilde{v}_0(t)\|_1\}).$$

By (7.46_k), $\Phi_9(|t|) \leq \Phi_{10}(|t|, \rho)$, where $\rho \equiv \max\left\{\|\vec{\Psi}\|_{X_0}, \|\vec{\tilde{\Psi}}\|_{X_0}\right\}$ and consequently

$$\frac{1}{2}\frac{\mathrm{d}}{\mathrm{d}t}\|\vec{w}(t)\|_{X_0}^2 + \nu\|w_1(t)\|^2 \leq \Phi_{10}(|t|, \rho)\|\vec{w}(t)\|_{X_0}^2. \tag{7.54}$$

The integration of this inequality yields

$$\|\vec{w}(t)\|_{X_0} \equiv \left\|V_t(\vec{\Psi}) - V_t(\vec{\tilde{\Psi}})\right\|_{X_0} \leq \Phi_{11}(|t|, \rho)\left\|\vec{\Psi} - \vec{\tilde{\Psi}}\right\|_{X_0} \tag{7.55}$$

for all $\vec{\Psi}, \vec{\tilde{\Psi}}$ belonging to the ball $B_\rho(0) \subset X_0$ of radius ρ. This inequality guarantees the uniform continuity of V_t on every ball $B_\rho(0)$. Since $V_t(\vec{\Psi})$ is continuous in t for each $\vec{\Psi} \in X_0$, $V_t(\vec{\Psi})$ is continuous in $(t, \vec{\Psi}) \in \mathbb{R} \times X_0$, i.e. the group $\{V_t, t \in \mathbb{R}, X_0\}$ is continuous.

Finally the boundedness of the semigroup $\{V_t, t \in \mathbb{R}^+, X_0\}$ follows from the estimate (7.46_1).

7.4 On differentiability of solution operators

Let us consider the differentiability of the solution operators V_t of problem (7.3). Assume that the conditions of Theorem 7.4 hold and, in addition, f is differentiable in the following sense:

(d) for all $u, \hat{u} \in H_1$

$$f(\hat{u}) - f(u) = f'(u)(\hat{u} - u) + r_f(\hat{u}, u) \tag{7.56}$$

where $f'(\hat{u}) \in L(H_1 \to H)$ and

$$\|f'(u)\|_{L(H_1\to H)} \leq \Phi_{12}(\|u\|_1); \tag{7.57}$$

the remainder $r_f(\hat{u}, u)$ satisfies the inequality

$$\|r_f(\hat{u}, u)\| \leq \Phi_{13}(\max\{\|\hat{u}\|_1, \|u\|_1\})\gamma(\|\hat{u} - u\|_1)\|\hat{u} - u\|_1 \tag{7.58}$$

where $\gamma : \mathbb{R}^+ \to \mathbb{R}^+$ is a continuous function, $\gamma(0) = 0$.

Due to hypothesis (a) (see (7.31)) condition (d) is a new restriction only for $\hat{u}$ close to u.

In this situation we shall prove that V_t is differentiable in the following sense:

$$V_t(\vec{\Psi} + s\vec{\xi}) - V_t(\vec{\Psi}) = sU(t, \vec{\Psi})\vec{\xi} + \vec{R}(t, s, \vec{\Psi}, \vec{\xi}) \tag{7.59}$$

for $\vec{\Psi} \in X_0$, $s \in [-s_0, s_0]$ and $\vec{\xi} \in X_0$ with $\|\vec{\xi}\|_{X_0} = 1$. Here $U(t, \vec{\Psi}) \in L(X_0 \to X_0)$ and $\mathbb{R}(\ldots)$ satisfy the inequality

$$\|\vec{R}(t, s, \vec{\Psi}, \vec{\xi})\|_{X_0} \le \Phi_{14}(|t|, |s|, \|\vec{\Psi}\|_{X_0}), \tag{7.60}$$

where $\Phi_{14}: \mathbb{R}^3 \to \mathbb{R}^+$ is a continuous nondecreasing function of its arguments and $|s|^{-1}\Phi_{14}(|t|, |s|, \|\vec{\Psi}\|_{X_0}) \to 0$ when $s \to 0$. The operator $U(t, \vec{\Psi}): X_0 \to X_0$ is the differential of V_t at the point $\vec{\Psi}$. We shall use the notation $\vec{v}(t, \vec{\Psi})$ for $V_t(\vec{\Psi})$, and $v_k(t, \vec{\Psi})$, $k = 0, 1$, for its components.

If the representation (7.59) really holds then it is easy to verify that $U(t, \vec{\Psi})\vec{\xi}$ is a solution of the linear problem

$$\partial_t \vec{u}(t) = a\vec{u}(t) + B(t, \vec{\Psi})\vec{u}(t), \quad \vec{u}\big|_{t=0} = \vec{\xi}, \tag{7.61}$$

where $B(t, \vec{\Psi})\vec{u}(t) = \begin{pmatrix} 0 \\ -f'(v(t, \vec{\Psi}))u_0(t) \end{pmatrix}$.

Formally, to obtain (7.61) one has to subtract from the equation (7.3) for $V_t(\vec{\Psi} + s\vec{\xi})$ the same equation written for $V_t(\vec{\Psi})$, to divide the result by s and then pass to the limit when $s \to 0$. Equation (7.61) is the so-called equation in variations for (7.3) (on the solution $v(t, \vec{\Psi})$).

It is easier to obtain the equation in variations for the scalar problem (7.1). This equation is

$$\partial_{tt}^2 u(t) + \nu\partial_t u(t) + Au(t) + f'(v_0(t, \Psi))\, u(t) = 0. \tag{7.61$'$}$$

The relations between $u(t)$ and $\vec{u}(t)$ are $u(t) = u_0(t)$, $\partial_t u(t) = u_1(t)$. Equation (7.61) is the vector form of equation (7.61′).

The hypothesis (7.57) implies that $B(t, \vec{\Psi})\vec{u} \in X_0$ for any $\vec{u} \in X_0$ and $\|f'(v_0(t, \vec{\Psi}))u_0\| \le \Phi_{12}(\|v_0(t, \vec{\Psi})\|_1)\|u_0\|_1$.

The unique solvability of problem (7.61) for any $\vec{\xi} \in X_0$ and $\vec{\Psi} \in X_0$ is proved as for problem (7.3) (see Theorem 7.4) but with essential simplifications due to the linearity of problem (7.61). Its solution $\vec{u}(t)$ enjoys the following properties:

$$\vec{u} \in C(\mathbb{R}, X_0), \qquad \mathrm{B}(t, \vec{\Psi})\vec{u}(t) \in X_0,$$
$$\|B(t, \vec{\Psi})\vec{u}(t)\|_{X_0} \le \Phi_{15}(|t|, \|\vec{\Psi}\|_{X_0}),$$
$$\partial_t \vec{u} \in L_{\infty, loc}(\mathbb{R}, X_{-1});$$

equation (7.61) is fulfilled in X_{-1} for all $t \in \mathbb{R}$ and the main energy relation

$$\frac{1}{2}\frac{\mathrm{d}}{\mathrm{d}t}\|\vec{u}(t)\|_{X_0}^2 = -\nu\|u_1(t)\|^2 - (f'(v_0(t, \vec{\Psi}))u_0(t), u_1(t)) \tag{7.62}$$

holds. From (7.62) and from the assumption (7.57) the estimate

$$\left\| U\left(t, \vec{\Psi}\right) \right\|_{L(X_0 \to X_0)} \leq \Phi_{16}\left(|t|, \|\vec{\Psi}\|_{X_0}\right) \tag{7.63}$$

is derived in the standard way.

Generally, $\Phi_{16} \to +\infty$ as $|t| \to +\infty$.

Thus we may take this fact for granted and estimate the remainder term

$$\vec{R}\left(t, s, \vec{\Psi}, \vec{\xi}\right) \equiv \vec{v}\left(t, \vec{\Psi} + s\vec{\xi}\right) - \vec{v}\left(t, \vec{\psi}\right) - s\vec{u}\left(t, \vec{\Psi}, \vec{\xi}\right),$$

where $\vec{u}\left(t, \vec{\Psi}, \vec{\xi}\right) \equiv U\left(t, \vec{\Psi}\right)\vec{\xi}$. It is clear that $\vec{R} \in C\left(\mathbb{R}, X_0\right)$. If we subtract from equation (7.3) for $\vec{v}\left(t, \vec{\Psi} + s\vec{\xi}\right)$ the same equation for $\vec{v}\left(t, \vec{\Psi}\right)$ and s times equation (7.61) for $\vec{u}\left(t, \vec{\Psi}, \vec{\xi}\right)$, we obtain for $\vec{R}\left(\ldots\right)$:

$$\partial_t \vec{R}\left(t, \ldots\right) = a\vec{R}\left(t, \ldots\right) + \vec{B}_1(t), \tag{7.64}$$

where $\vec{B}_1(t) = \begin{pmatrix} 0 \\ B_1(t,\ldots) \end{pmatrix}$ and

$$\begin{aligned} B_1(t, \ldots) \equiv & -f\left(v_0\left(t, \vec{\Psi} + s\vec{\xi}\right)\right) + f\left(v_0\left(t, \vec{\Psi}\right)\right) \\ & + sf'\left(v_0\left(t, \vec{\Psi}\right)\right)u_0\left(t, \vec{\Psi}, \vec{\xi}\right). \end{aligned}$$

Moreover, it is evident that

$$\vec{R}|_{t=0} = 0. \tag{7.65}$$

From the estimates (7.46_1) and (7.63) and the assumption (7.57) we deduce that $B_1 \in L_{\infty,loc}\left(\mathbb{R}, H\right)$. The vector-valued function $\vec{\mathbb{R}}: t \to X_0$ may be regarded as a solution (from $C\left(\mathbb{R}, X_0\right)$) of the linear problem (7.11) with the source term $\vec{g}(t) = \vec{B}_1(t)$ and with zero initial value.

According to Theorem 7.2 the following energy relation holds:

$$\frac{1}{2}\frac{\mathrm{d}}{\mathrm{d}t}\|\vec{R}\left(t, \ldots\right)\|^2_{X_0} = -\nu\|R_1\left(t, \ldots\right)\|^2 + \left(B_1\left(t, \ldots\right), R_1\left(t, \ldots\right)\right). \tag{7.66}$$

$B_1(t, \ldots)$ using (7.56) may be represented in the form

$$\begin{aligned} B_1(t, \ldots) = & -f'\left(v_0\left(t, \vec{\Psi}\right)\right)R_0\left(t, s, \vec{\Psi}, \vec{\xi}\right) \\ & - r_f\left(v_0\left(t, \vec{\Psi} + \vec{s}\xi\right), v_0\left(t, \vec{\Psi}\right)\right), \end{aligned}$$

where $R_0(t, \ldots)$ is the first component of the vector $\vec{R}(t, \ldots)$; note that the second component $R_1\left(t, \ldots\right)$ appeared in (7.66). The last term in (7.66) may be estimated as follows:

$$\begin{aligned}
&\left|\left(B_1(t,\ldots), R_1(t,\ldots)\right)\right| \\
&\quad \le \Phi_{12}\left(\left\|v_0(t,\vec{\Psi})\right\|_1\right)\|R_0(t,\ldots)\|_1\|R_1(t,\ldots)\| \\
&\qquad + \Phi_{13}\left(\max\left\{\left\|v_0(t,\vec{\Psi}+s\vec{\xi})\right\|_1, \left\|v_0(t,\vec{\Psi})\right\|_1\right\}\right) \\
&\qquad \cdot \gamma\left(\left\|v_0(t,\vec{\Psi}+s\vec{\xi}) - v_0(t,\vec{\Psi})\right\|_1\right) \\
&\qquad \cdot \left\|v_0(t,\vec{\Psi}+s\vec{\xi}) - v_0(t,\vec{\Psi})\right\|_1 \|R_1(t,\ldots)\| \\
&\quad \le \Phi_{17}\left(|t|, \|\vec{\Psi}\|_{X_0}\right)\|\vec{R}(t,\ldots)\|_{X_0}^2 \\
&\qquad + \Phi_{18}\left(|t|, |s|, \|\vec{\Psi}\|_{X_0}\right)\|\vec{R}(t,\ldots)\|_{X_0},
\end{aligned} \tag{7.67}$$

where $\Phi_{18}(\ldots)$ is such that $|s|^{-1}\Phi_{18}\left(|t|, |s|, \|\vec{\Psi}\|_{X_0}\right) \to 0$ when $s \to 0$. We have used the assumptions (7.57), (7.58), the estimates (7.46_k) for the solutions $\vec{v}(t, \vec{\Psi}+s\vec{\xi})$ and $\vec{v}(t, \vec{\Psi})$, and the estimate (7.55) for their difference. Substituting (7.67) into (7.66) and integrating the resulting inequality, keeping in mind (7.65), we obtain the desired estimate (7.60). Let us summarize the obtained results in the following theorem.

Theorem 7.5 *Let the assumptions of Theorem 7.4 and the condition (d) be fulfilled. Then for all $t \in \mathbb{R}$ the operator V_t is differentiable at each point $\vec{\Psi} \in X_0$, its differential $U(t, \vec{\Psi})$ belongs to the class $L(X_0 \to X_0)$ and the norms $\left\|U(t,\vec{\Psi})\right\|_{L(X_0\to X_0)}$ are uniformly bounded for $\vec{\Psi} \in B_\rho(0) \subset X_0$, $\forall_\rho > 0$. The estimates (7.60) hold.*

7.5 On the affiliation of the semigroup $\{V_t, t \in \mathbb{R}^+, X_0\}$ to the class $\mathcal{AK}$ and attractors

We shall prove the following

Theorem 7.6 *Let the conditions of Theorem 7.4 and the following condition be satisfied*

(e) At every point of $u \in H_1$ f has a derivative $f'(u) \in L(H \to H_{-1+\delta})$ where δ is a number belonging to $(0, 1]$ and

$$\|f'(u)\|_{L(H\to H_{-1+\delta})} \le \Phi_{19}(\|u\|_1). \tag{7.68}$$

Then the semigroup $\{V_t, t \in \mathbb{R}^+, X_0\}$ belongs to the class $\mathcal{AK}$.

Let $\vec{v}(t, \vec{\Psi})$ be a solution of problem (7.3) guaranteed by Theorem 7.4. Its representation (7.37) holds and it may be interpreted as the representation of the operators $V_t : X_0 \to X_0$ in the form

$$V_t = U_t + W_t, \quad t \in \mathbb{R}^+, \tag{7.69}$$

where $U_t : X_0 \to X_0$ are the resolving operators of the linear problem (7.22) and $W_t : X_0 \to X_0$ are defined by the equality

$$W_t(\vec{\Psi}) = \int_0^t U_{t-\tau}\begin{pmatrix} 0 \\ -f(v_0(\tau, \vec{\Psi})) + h \end{pmatrix} d\tau. \tag{7.70}$$

Let us show that (7.69) allows us to use Theorem 3.3, and thus to prove that the semigroup $\{V_t, t \in \mathbb{R}^+, X_0\}$ belongs to the class $\mathcal{AK}$.

We already know that $\|U_t\|_{L(X_0 \to X_0)} \to 0$, when $t \to +\infty$; therefore we only have to prove the compactness of the operators W_t, $t > 0$. To this end it is sufficient to prove that each W_t $(t > 0)$ transforms any set B, bounded in X_0, into a set bounded in X_δ, $\delta > 0$. We also know that

$$\vec{v}(\cdot\,, \vec{\Psi}) \in C(R^+, X_0) \quad \text{and}$$
$$\sup_{t \in R^+} \|\vec{v}(t, \vec{\Psi})\|_{X_0} \le \Phi_{20}(\|\vec{\Psi}\|_{X_0}),$$

where $\Phi_{20}(\|\vec{\Psi}\|_{X_0}) = \left(\max\{1; \nu_1^{-1}\}\Phi_4(\|\vec{\Psi}\|_{X_0})\right)^{1/2}$ (see (7.45)).

From this and from condition (a) it follows that

$$g(\cdot\,, \vec{\Psi}) \equiv -f(v_0(\cdot\,, \vec{\Psi})) + h$$

belongs to $C(\mathbb{R}^+, H)$ and

$$\sup_{t \in \mathbb{R}^+} \|g(t, \vec{\Psi})\| \le \sup_{t \in \mathbb{R}^+} [\Phi_1(\|v_0(t, \vec{\Psi})\|_1)\|v_0(t, \vec{\Psi})\|_1 + \|h\|].$$

Additionally, because of condition (e), $g(t, \vec{\Psi})$ is differentiable in t,

$$\begin{aligned} \partial_t g(t, \vec{\Psi}) &= -f'(v_0(t, \vec{\Psi}))\partial_t v_0(t, \vec{\Psi}) \\ &= -f'(v_0(t, \vec{\Psi}))v_1(t, \vec{\Psi}) \in H_{-1+\delta} \end{aligned}$$

and

$$\|\partial_t g(t, \vec{\Psi})\|_{-1+\delta} \le \Phi_{19}(\|v_0(t, \vec{\Psi})\|_1)\|v_1(t, \vec{\Psi})\|.$$

Therefore $\sup_{t \in \mathbb{R}^+} \|\partial_t g(t, \vec{\Psi})\|_{-1+\delta} \le \Phi_{21}(\|\vec{\Psi}\|_{X_0})$.

We can consider $\vec{W}_t(\vec{\Psi}) \equiv \vec{w}(t, \vec{\Psi})$ as a solution of the linear problem (7.11) with the free term $g(t, \vec{\Psi})$ and zero initial data and use Theorem 7.3. All its conditions are fulfilled when $r = -1 + \delta$, hence

$$w_0(\cdot, \vec{\Psi}) \in C(R, H_{1+\delta}),\ w_1(\cdot, \vec{\Psi}) = \partial_t w_0(\cdot, \vec{\Psi}) \in C(\mathbb{R}, H_\delta)$$

and, as is easy to verify, from (7.28) and (7.29) we get the estimate

$$\sup_{t \in R^+} \|W_t(\vec{\Psi})\|_{X_\delta} \leq \Phi_{22}(\|\vec{\Psi}\|_{X_0}), \quad \delta > 0. \tag{7.71}$$

Thus Theorem 7.6 is proved as X_δ with $\delta > 0$ is compactly embedded in X_0.

Now we shall use the theorems of Part I on attractors for semigroups of the class $A\mathcal{K}$; see Chapter 3. To this end we must prove at least the point-dissipativeness of our semigroup.

Since, in this case, we have the good Lyapunov function $\mathcal{L}$ (see (7.39)) it is sufficient to prove the boundedness of the set Z of all possible stationary points $\vec{z} = \binom{z_0}{z_1}$. Here $z_1 = 0$ and z_0 is a solution of problem

$$Az_0 + f(z_0) = h, \quad z_0 \in H_1. \tag{7.72}$$

Let us assume that $f(\cdot)$ has the property:

(f) for any $u \in H_1$

$$-(f(u), u) \leq (1 - \nu_2)\|u\|_1^2 + c_2,\ \nu_2 \in (0, 1],\ \ c_2 \in \mathbb{R}^+. \tag{7.73}$$

The inner product in H of (7.72) with z_0 and (7.73) give

$$\|z_0\|_1^2 \leq (1 - \nu_2)\|z_0\|_1^2 + c_2 + \|h\|\|z_0\|;$$

hence the estimate

$$\|z_0\|_1 \leq c_3 \tag{7.74}$$

holds for any possible solution of problem (7.72).

This estimate and the hypothesis (a) on f allow us also to majorize the norm $\|z_0\|_2$, namely

$$\begin{aligned}\|z_0\|_2 = \|Az_0\| &\leq \|f(z_0)\| + \|h\| \\ &\leq \Phi_1(\|z_0\|_1)\|z_0\|_1 + \|h\| \\ &\leq \Phi_1(c_3)c_3 + \|h\| \equiv c_4.\end{aligned} \tag{7.75}$$

Thus the set Z lies in the ball $B_{c_3}(0) \subset X_0$ and is bounded in X_1.

Moreover, Z is closed in X_0 (and in X_1 as well, see 7.31) and therefore it is compact in X_0.

So, if the conditions of Theorem 7.6 and hypothesis (7.73) are fulfilled, then, according to Theorem 7.2, $\{V_t, t \in \mathbb{R}^+, X_0\}$ has the set Z as its attractor $\widehat{\mathcal{M}}$ and the attractor $\mathcal{M}$ is a connected compact set, composed of Z and of all complete trajectories connecting the points of Z. Moreover, the attractor $\mathcal{M}$ is a closed bounded subset of the space X_δ, $\delta > 0$. Indeed, if $\vec{\Psi} \in \mathcal{M}$, then $\vec{v}(t, \vec{\Psi}) \equiv V_t(\vec{\Psi}) \in \mathcal{M}$, for every $t \in \mathbb{R}$, and for arbitrary t and s we have

$$\vec{v}(t, \vec{\Psi}) = U_{t-s}(\vec{v}(s, \vec{\Psi})) + \int_s^t U_{t-\tau}\begin{pmatrix} 0 \\ -f(v_0(\tau, \vec{\Psi})) + h \end{pmatrix} d\tau. \tag{7.76}$$

If s tends to $-\infty$, then we get

$$\vec{v}(t, \vec{\Psi}) = \int_{-\infty}^t U_{t-\tau}\begin{pmatrix} 0 \\ -f(v_0(\tau, \vec{\Psi})) + h \end{pmatrix} d\tau, \; t \in R, \tag{7.77}$$

since $\mathcal{M}$ is a bounded set and the operators U_t enjoy the property (7.23_1).

This is the integral equation for complete trajectories lying in $\mathcal{M}$. For $\mathcal{M}$ the following estimate holds

$$\|\mathcal{M}\|_{X_\delta} \leq \Phi_{22}\left(\|\mathcal{M}\|_{X_0}\right), \tag{7.78}$$

where $\|\mathcal{M}\|_X := \sup_{\vec{\Psi} \in \mathcal{M}} \|\vec{\Psi}\|_X$.

Indeed, let $\vec{\Psi} \in \mathcal{M}$. We can use (7.69) for V_s with $s > 0$ and represent $\vec{\psi}$ as follows:

$$\vec{\Psi} = V_s(\vec{v}(-s, \vec{\Psi})) = U_s(\vec{v}(-s, \vec{\Psi})) + W_s(\vec{v}(-s, \vec{\Psi})),$$

where $\vec{v}(-s, \vec{\Psi}) = V_{-s}(\vec{\Psi}) \in \mathcal{M}$.

Then $\sup_{s \in \mathbb{R}^+} \|\vec{v}(-s, \vec{\Psi})\|_{X_\delta} \leq \|\mathcal{M}\|_{X_\delta}$, because $\vec{v}\,(-s, \vec{\Psi}) \in \mathcal{M}$, and by ($3.23_1$) $\|U_s(\vec{v}(-s, \vec{\Psi}))\|_{X_\delta} \leq m_4(\alpha, \delta)e^{-\alpha s}\|\mathcal{M}\|_{X_\delta}$. For the second term $W_s(\vec{v}(-s, \vec{\Psi}))$ we have the estimate (7.71) and therefore

$$\|\vec{\Psi}\|_{X_\delta} \leq m_4(\alpha, \delta)e^{-\alpha s}\|\mathcal{M}\|_{X_\delta} + \Phi_{22}(\|\mathcal{M}\|_{X_0}).$$

If s tends to $+\infty$ then we get (7.78).

So we have proved the following theorem:

Theorem 7.7 *If $h \in H$ and f satisfies the conditions (a)–(c), (e) and (f) (see (7.73)), then the semigroup $\{V_t, t \in \mathbb{R}^+, X_0\}$ has the compact minimal global attractor $\widehat{\mathcal{M}}$ coinciding with the set Z of all stationary points. It is a bounded subset in the space X_1. Its minimal global B-attractor $\mathcal{M}$ is connected and* *compact in X_0, consisting of all complete trajectories connecting points of the set Z. It is a bounded subset in the space X_δ, $\delta > 0$. For the trajectories $\vec{v}(t, \vec{\Psi}) = V_t(\vec{\Psi})$ on $\mathcal{M}$ the integral equation (7.77) holds.*

Although Theorem 7.7 is meaningful, its practical application for determining $\mathcal{M}$ requires a lot of additional work.

That is why those cases where we can really find a bounded set B_0 containing the attractor $\mathcal{M}$ are very important. Now we do this by supposing that $\mathcal{F}$, besides the properties listed above, also enjoys the following:

(g) for all $u \in H_1$

$$\nu_3 \mathcal{F}(u) - (f(u), u) \leq c_3, \tag{7.79}$$

with some $\nu_3 \in (0, 2)$ and $c_3 \in \mathbb{R}^+$.

As in the proof of the estimates (7.20_k) we shall use $\vec{v}(t, \alpha)$ related to the solution $v(t)$ by means of (7.13); $\vec{v}(t, \alpha)$is the solution of (7.4) and hence the corresponding energy equality of the form (7.16) with $s = 0$ holds. Namely,

$$\begin{aligned} \frac{1}{2}\frac{\mathrm{d}}{\mathrm{d}t}\|\vec{v}(t, \alpha)\|^2_{X_0} = -\alpha\|\vec{v}_0(t, \alpha)\|^2_{1, \alpha} - (\nu - \alpha)\|v_1(t, \alpha)\|^2 \\ + (-f(v_0(t, \alpha)) + h, v_1(t, \alpha)). \end{aligned} \tag{7.80}$$

The last term may be transformed as follows:

$$\begin{aligned} &(-f(v_0(t, \alpha)) + h, \alpha v_0(t, \alpha) + \partial_t v_0(t, \alpha)) \\ &\quad = \frac{\mathrm{d}}{\mathrm{d}t}[-\mathcal{F}(v_0(t, \alpha)) + (h, v_0(t, \alpha))] \\ &\qquad - \alpha(f(v_0(t, \alpha)), v_0(t, \alpha)) + \alpha(h, v_0(t, \alpha)). \end{aligned}$$

This enables us to rewrite (7.80) in terms of the function

$$\mathcal{L}\alpha(\vec{v}) \equiv \frac{1}{2}\|u_0\|^2_{1, \alpha} + \frac{1}{2}\|u_1\|^2 + \mathcal{F}(u_0) - (h, u_0).$$

Namely

$$\begin{aligned} &\frac{\mathrm{d}}{\mathrm{d}t}L_\alpha(\vec{v}(t, \alpha)) \\ &\quad = -\alpha\|v_0(t, \alpha)\|^2_{1, \alpha} - (\nu - \alpha)\|v_1(t, \alpha)\|^2 \\ &\qquad - \alpha(f(v_0(t, \alpha)), v_0(t, \alpha)) + \alpha(h, v_0(t, \alpha)) \end{aligned}$$

$$
\begin{aligned}
&= -\alpha \nu_3 L_\alpha(\vec{v}(t,\alpha)) \\
&\quad + \alpha\left[\nu_3 \mathcal{F}(v_0(t,\alpha)) - (f(v_0(t,\alpha)), v_0(t,\alpha))\right] \\
&\quad + \alpha(1-\nu_3)(h, v_0(t,\alpha)) \\
&\quad - \alpha\left(1 - \frac{1}{2}\nu_2\right)\|v_0(t,\alpha)\|^2_{1,\alpha} - \left(\nu - \alpha - \frac{1}{2}\alpha\nu_3\right)\|v_1(t,\alpha)\|^2.
\end{aligned}
\tag{7.81}
$$

By the assumption (7.79) the term enclosed in the square brackets [...] on the right hand side of (7.81) is no larger than αc_3. The sum of the remaining terms (except the first one) is no larger than αc_4, where

$$
c_4 = [(2-\nu_3)\lambda_1(\alpha)]^{-1}(1-\nu_3)^2\|h\|,
$$

provided α satisfies the inequality (7.6). Now it follows from (7.81) that

$$
\frac{\mathrm{d}}{\mathrm{d}t}\mathcal{L}_\alpha(\vec{v}(t,\alpha)) \le -\alpha\nu_3\mathcal{L}_\alpha(\vec{v}(t,\alpha)) + \alpha c_5, \tag{7.82}
$$

with $c_5 = c_3 + c_4$. By integrating this inequality we obtain

$$
\mathcal{L}_\alpha(\vec{v}(t,\alpha)) \le e^{-\alpha\nu_3 t}\mathcal{L}_\alpha(\vec{v}(0,\alpha)) + \nu_3^{-1}c_5,\ t \in \mathbb{R}^+. \tag{7.83}
$$

From (7.83) we want to deduce information about the solution $\vec{v}(t) = C^{-1}(\alpha)\vec{v}(t,\alpha)$ in terms of the norm in X_0. We proceed as above, where the estimates (7.20_k) were deduced from (7.18_k). Namely, for arbitrary $\vec{v}$, $\vec{v}(\alpha) \in X_0$ such that $\vec{v}(\alpha) = C(\alpha)\vec{v}$ we have (7.19_2) with $s = 0$, i.e.

$$
m_1^{-1}(0,\alpha)\|\vec{v}\|_{X_0} \le \|\vec{v}(\alpha)\|_{X_{0,\alpha}} \le m_2(0,\alpha)\|\vec{v}\|_{X_0}. \tag{7.84}
$$

Using (7.84) and (7.36) we get

$$
\begin{aligned}
\mathcal{L}_\alpha(\vec{v}(\alpha)) &\le \frac{1}{2}m_2^2(0,\alpha)\|\vec{v}\|^2_{X_0} + \Phi_3(\|\vec{v}\|_{X_0}) + \|h\|\lambda_1^{-1/2}\|\vec{v}\|_{X_0} \\
&= \Phi_{23}(\|\vec{v}\|_{X_0}).
\end{aligned}
\tag{7.85}
$$

In order to estimate $\mathcal{L}_\alpha(\vec{v}(\alpha))$ from below we use (7.84) and (7.34):

$$
\begin{aligned}
\mathcal{L}_\alpha(\vec{v}(\alpha)) \ge \frac{1}{2}m_1^{-2}(0,\alpha)\|\vec{v}\|^2_{X_0} - \left(\frac{1}{2} - \nu_1\right)\|\vec{v}\|^2_{X_0} \\
- c_1 - \|h\|\lambda_1^{-1/2}\|\vec{v}\|_{X_0}.
\end{aligned}
\tag{7.86}
$$

In addition to (7.6) we impose on α the following restriction

$$
\begin{aligned}
m_1^{-2}(0,\alpha) - 1 + 2\nu_1 &= (\mu_1(\alpha)m^{-1}(\alpha))^2 - 1 + 2\nu_1 \\
&\ge 4\nu_4 > 0
\end{aligned}
\tag{7.87}
$$

with some positive ν_4. Since $\mu_1(\alpha)$ and $m(\alpha)$ tend to 1 as $\alpha \to 0$, (7.87) always holds for α small enough. Now, (7.86) and (7.87) yield

$$\begin{aligned}\mathcal{L}_\alpha(\vec{v}(\alpha)) &\geq \nu_4\|\vec{v}\|^2_{X_0} - c_1 - (4\nu_4\lambda_1)^{-1}\|h\|^2 \\ &\equiv \nu_4\|\vec{v}\|^2_{X_0} - c_5'.\end{aligned} \tag{7.88}$$

From the inequalities (7.83), (7.85) and (7.88) it follows that

$$\|\vec{v}(t, \vec{\Psi})\|^2_{X_0} \leq e^{-\alpha\nu_3 t}\nu_4^{-1}\Phi_{23}(\|\vec{\Psi}\|_{X_0}) + c_6^2,\ t \in \mathbb{R}^+, \tag{7.89}$$

where $c_6^2 = \nu_4^{-1}(c_5' + \nu_3^{-1}c_5)$. This estimate, being true for a solution $\vec{v}(t, \vec{\Psi})$ of (7.3) with arbitrary $\vec{\Psi} \in X_0$, shows that every bounded set $B \subset X_0$ falls inside the ball

$$\mathrm{B}_{2c_6}(0) \equiv \{\vec{v} \colon \vec{v} \in X_0,\ \|\vec{v}\|_{X_0} \leq 2c_6\} \tag{7.90}$$

after a finite time $t(B)$. Thus we have proved:

Theorem 7.8 *If $h \in H$ and f satisfies the conditions (a)–(c) and (g), then $B_{2c_6}(0)$ is an absorbing set of the semigroup $\{V_t,\ t \in \mathbb{R}^+, X_0\}$.*

Remark A representative class of nonlinear partial differential equations is that in which $f(v)$ is a polynomial $P_m(v)$ of degree m. If m is odd and the coefficient of the principal term in $P_m(v)$ is positive, the conditions (7.34) and (7.79) hold with $\nu_1 = \frac{1}{2}$ (7.34) and $\nu_4 = \frac{1}{m+1}$ (7.79).

7.6 On the dimensions of compact invariant sets

We want here to estimate above $\dim_H(\mathcal{A})$ and $\dim_f(\mathcal{A})$ for a compact invariant set $\mathcal{A} \in X_0$ using Theorems 4.8 and 4.9. Let f satisfy the conditions of Theorem 7.5, its derivative $f'(v_0)$ at each point $v_0 \in \mathcal{P}\mathcal{A} \subset H_1$, where $\mathcal{P}$ is the orthoprojector of X_0 onto H_1, belongs to $L(H_{1-\beta} \to H_\beta)$ with some $\beta > 0$ and

$$\sup_{v_0 \in \mathcal{P}\mathcal{A}} \|f'(v_0)\|_{L(H_{1-\beta} \to H_\beta)} \leq c_1, \quad c_1 \in \mathbb{R}^+. \tag{7.91}$$

We shall use the representation of problem (7.1) in the form (7.4) and $X_{0,\alpha}$ as the phase-space, supposing that α is subject to the restrictions (7.6).

The next inequality follows from (7.91):

$$\sup_{v_0 \in \mathcal{P}(\alpha)\mathcal{A}} \|f'(v_0)\|_{L(H_{1-\beta,\alpha} \to H_{\beta,\alpha})} \leq c_1(\alpha), \tag{7.92}$$

where $\mathcal{P}(\alpha)$ is the orthoprojector of $X_{0,\alpha}$ onto $H_{1,\alpha}$.

Let us denote by $V_t(\alpha)\colon X_{0,\alpha} \to X_{0,\alpha}$ the solution operators of problem (7.4). It is easy to verify that they have the same properties as those $V_t\colon X_0 \to X_0$ of problem (7.3). Denote by $U(t, \alpha, \vec{\Psi})$ the differential of $V_t(\alpha)$ at the point $\vec{\Psi} \in \mathcal{A}$. The family $\{U(t, \alpha, \vec{\Psi})\colon X_{0,\alpha} \to X_{0,\alpha}, t \in \mathbb{R}^+\}$ is a collection of solution operators of the linear problem

$$\begin{aligned} \partial_t \vec{u}(t, \alpha) &= a(\alpha)\vec{u}(t, \alpha) + B(t, \alpha, \vec{\Psi})\vec{u}(t, \alpha) \\ &\equiv L(t, \alpha, \vec{\Psi})\vec{u}(t, \alpha) \\ \vec{u}|_{t=0} &= \vec{\xi}, \end{aligned} \tag{7.93}$$

where $B(t, \alpha, \vec{\Psi})\vec{u}(t, \alpha) = \begin{pmatrix} 0 \\ -f'(v_0(t,\alpha,\vec{\Psi}))u_0(t, \alpha) \end{pmatrix}$ (see (7.61)).

According to Theorems 4.8 and 4.9 we have to majorize the quadratic form $(L(t, \alpha, \vec{\Psi})\vec{u}, \vec{u})_{X_{0,\alpha}}$ for an arbitrary $\vec{u} \in X_{0,\alpha}$. From (7.10) and (7.91), we get (as in (7.16)):

$$\begin{aligned} &((L(t, \alpha, \vec{\Psi})\vec{u}, \vec{u})_{X_{0,\alpha}} \\ &\quad\le -\alpha\|u_0\|^2_{1,\alpha} - (\nu - \alpha)\|u_1\|^2 - (f'(v_0(t, \alpha, \vec{\Psi}))u_0, u_1) \\ &\quad\le -\alpha\|\vec{u}\|^2_{X_{0,\alpha}} + \|f'(v_0(t, \alpha, \vec{\Psi}))u_0\|_{\beta,\alpha}\|u_1\|_{-\beta,\alpha} \\ &\quad\le -\alpha\|\vec{u}\|^2_{X_{0,\alpha}} + c_1(\alpha)\|u_0\|_{1-\beta,\alpha}\|u_1\|_{-\beta,\alpha} \\ &\quad\le -\alpha\|\vec{u}\|^2_{X_{0,\alpha}} + c_2(\alpha)\|\vec{u}\|^2_{X_{-\beta,\alpha}}, \end{aligned} \tag{7.94}$$

where $c_2(\alpha) = \frac{1}{2}c_1(\alpha)$. Thus, the condition (4.34) of Theorem 4.8 is fulfilled with $h_0(t) = \alpha$, $m = 1$ and $h_{s_1}(t) = c_2(\alpha)$, $s_1 = -\beta$. Therefore $\dim_H(\mathcal{A}) \le N$ where N is the minimal integer satisfying the inequality

$$-\alpha N + c_2(\alpha)\mathrm{Sp}_N\, \vec{A}^{-\beta}(\alpha) < 0. \tag{7.95}$$

Let us remember that

$$\vec{A}(\alpha) := \begin{pmatrix} A(\alpha) & 0 \\ 0 & A(\alpha) \end{pmatrix} \quad A(\alpha) = A - \alpha(\nu - \alpha)I.$$

The spectrum of $\vec{A}(\alpha)$ consists of the numbers $\lambda^{\pm}_\kappa(\alpha) = \lambda_\kappa(\alpha) = \lambda_\kappa - \alpha(\nu - \alpha)$, $k = 1, 2, \ldots$ where $0 < \lambda_1 \le \lambda_2 \le \cdots$ are eigenvalues of A.

In order to estimate $\dim_f(\mathcal{A})$ we suppose that

$$\mathrm{Sp}_n\, \vec{A}^{-\beta}(\alpha) \le c_3(\alpha)n^{1-\beta}, \quad n = 1, 2, \ldots \tag{7.96}$$

with some $c_3(\alpha) \in \mathbb{R}^+$ Then, according to Theorem 4.9 $\dim_f(\mathcal{A}) \le N$, where N is the minimal integer such that

$$-\alpha + c_2(\alpha)c_3(\alpha)N^{-\beta} < 0. \tag{7.97}$$

We know that $X_{0,\alpha}$ and X_0 coincide as sets and their norms are equivalent. Therefore the dimensions $\dim_H(\cdot)$ and $\dim_f(\cdot)$ of $\mathcal{A}$ as subsets of the space $X_{0,\alpha}$ and of the space X_0 are the same. So we have proved the following theorem:

Theorem 7.9 *Let the conditions of Theorem 7.5 and condition* (7.91) *be fulfilled, and* α *be a number satisfying the inequalities* (7.6). *Then* $\dim_H(\mathcal{A}) \leq N$ *where* N *is the minimal integer satisfying the inequality* (7.95). *If the spectrum* $\{\lambda_k\}_{k=1}^{\infty}$ *of operator* A *is such that the inequality* (7.96) *is satisfied, then* $\dim_f(\mathcal{A}) \leq N$ *where* N *is the minimal integer satisfying* (7.97).

References

[1] Ladyzhenskaya, O. On the dynamical system generated by the Navier–Stokes equations. *Zapiski nauchnykh seminarov LOMI. Leningrad.* **27** (1972) 91–114 (English translation in *J. of Soviet Math.* **3** 4 (1975))

[2] Mallet-Paret, J. Negatively invariant sets of compact maps and an extension of a theorem of Cartwright. *J. Diff. Eq.* **22** (1976) 331–48

[3] Douady A., & Oesterlé J. Dimension de Hausdorff des attracteurs. *C. R. Acad. Sc.* **290** (1980) 1135–38

[4] Foias, C., & Temam, R. Some analytic and geometric properties of the solutions of the Navier–Stokes equations. *J. Math. Pures Appl.* **58**, fasc.3 (1979) 339–68

[5] Il'ashenko, U. S. Weak contractive systems and attractors for Galerkin's approximations of the Navier–Stokes equations. *Uspechi Math. Nauk.* **36**, 3 (1981) 243–4

[6] Ladyzhenskaya, O. On finite dimensionality of bounded invariant sets for the Navier–Stokes equations and some other dissipative systems. *Zapiski nauchnykh seminarov LOMI* **115** (1982) 137–155

[7] Babin A. V., Vishik M. I. Attractors for the Navier–Stokes system and parabolic equations and an estimation of its dimension. *Zapiski nauchnykh seminarov LOMI* **115** (1982) 3–15

[8] Il'ashenko, U. S. On the dimension of attractors for k-contractive systems in an infinite-dimensional space. *Vestnik MGU. ser. math, and mech.* **3** (1983)

[9] Constantin P., Foias C., & Temam R. *Attractors representing turbulent flows. American Mathematical Society, (1985).*

[10] Ghidaglia J. M., & Temam R. Attractors for damped nonlinear hyperbolic eqiatons. *J. Math. Pures Appl.* **66**, 3 (1987) 273–319.

[11] Ladyzhenskaya, O. On estimates of the fractal dimension and the number of determining modes for invariant sets of dynamical systems. *Zapiskii nauchnich seminarovs LOMI* **163** (1987) 105–29

[12] Ladyzhenskaya, O. On finding the minimal global attractors for the Navier–Stokes equations and other PDE. *Uspechi Math. Nauk.* **42**, n. 6 (1987) 25–60

[13] Hale, J. K. *Theory of functional-differential equations. Springer-verlag, Berlin–Heidelberg–New York,* (1977)

[14] Henry, D. *Geometric theory of semi-linear parabolic equations. Springer-Verlag, Berlin–Heidelberg–New York,* (1981)

[15] Ladyzhenskaya O., Solonnikov V., & Uralcéva N. *Linear and quasilinear equations of parabolic type. M., Nauka* (1967)

[16] Ladyzhenskaya O., & Uralcéva N. Survey of results on solvability of boundary value problems for uniformly elliptic and parabolic quasilinear equations of second order with unbounded singularities. *Uspechi Math. Nauk.* **41**, n. 5 (1986) 59–83

Index

Printed in the United States
by Baker & Taylor Publisher Services